"The embrace of markets as the means by which urban life can and should be monitored, fixed and managed never went away. Neoliberalism remains a foundational component of the imaginaries of economic and urban governance institutions. This book shows why this matters to the everyday life of cities and the people living within these increasingly fractured and unequal domains. Taking as its terrain an urban world in the midst of social, political and economic crisis this book deliberates on how this is also in many ways an urban crisis. A diverse set of empirically grounded contributions covers the key dimensions of domestic life, processes of racial banishment, eviction, foreclosure, protest, inequality, segregation and the neglect of marginal populations. This is an important and timely collection that will be consulted widely by urbanists globally."

Rowland Atkinson, *Professor and Research Chair in Inclusive Societies, Department of Urban Studies and Planning, University of Sheffield, UK*

"This volume offers a crucial set of concepts and tools to help scholars and other readers make sense of contemporary dynamics in housing and urban society. Examining cities across the world, the book systematically analyses the neoliberal assault on urban life. In its explorations of the politics and experience of dwelling in the neoliberal city, the book helps us think critically about one of the central challenges of our times."

David J. Madden, *Associate Professor in Sociology and Co-Director of the Cities Programme, London School of Economics and Political Science, UK*

Urban Change and Citizenship in Times of Crisis

The contributions to *Urban neo-liberalisation* bring together critical analyses of the dynamics and processes neo-liberalism has facilitated in urban contexts. Recent developments, such as intensified economic investment and exposure to aggressive strategies of banks, hedge-funds and investors, and long-term processes of market- and state-led urban restructuration, have produced uneven urban geographies and new forms of exclusion and marginality. These strategies have no less transformed the governance of cities by subordinating urban social life to rationalities and practices of competition within and between cities, and they also heavily impact on city inhabitants' experience of everyday life. Against the backdrop of recent austerity politics and a marketisation of cities, this volume discusses processes of urban neo-liberalisation with regard to democracy and citizenship, inclusion and exclusion, opportunities, and life-chances. It addresses pressing issues of commodification of housing and home, activation of civil society, vulnerability, and the right to the city.

Bryan S. Turner is Professor of Sociology and Director of the Institute for Religion Politics and Society at the Australian Catholic University, Honorary Professor and Director of the Centre for Citizenship, Social Pluralism and Religious Diversity at Potsdam University, Germany, and Emeritus Professor at the Graduate Center at the City University of New York City. He is the founding editor of the *Journal of Classical Sociology*. He edited the *Blackwell Wiley Encyclopedia for Social Theory* (2018). He was awarded a Doctor of Letters by Cambridge University in 2009 and received the Max Planck Award in social science in 2015.

Hannah Wolf is a researcher and lecturer completing her PhD at the University of Potsdam, Coordinator of the Centre for Citizenship, Social Pluralism, and Religious Diversity and associate member at the collaborative DFG-research centre Re-Figuration of Spaces (TU Berlin). Her academic background includes theatre and media studies, anthropology, philosophy, and sociology. Her research interests lie in the political and moral economies of housing and home, citizenship, urban sociology, and the sociology of everyday life.

Gregor Fitzi is Co-Director of the Centre for Citizenship, Social Pluralism and Religious Diversity at University of Potsdam, Germany. After his PhD in Sociology at the University of Bielefeld, he was Assistant Professor at the Institute of Sociology, University of Heidelberg, Germany. Among his recent publications are *The Challenge of Modernity. Simmel's Sociological Theory* (2019) and *Populism and the Crisis of Democracy*, 3 vols. edited with Jürgen Mackert and Bryan S. Turner (2019).

Jürgen Mackert is Professor of Sociology and a Co-Director of the Centre for Citizenship, Social Pluralism, and Religious Pluralism at the University of Potsdam, Germany. His research interests lie in the sociology of citizenship, political economy, closure theory, (neo-)liberalism, settler colonialism and de-democratisation. His most recent publication is *Populism and the Crisis of Democracy*, 3 vols., edited with Gregor Fitzi and Bryan S. Turner (2019).

Urban Change and Citizenship in Times of Crisis
Volume 1: Theories and Concepts
Edited by Bryan S. Turner, Hannah Wolf, Gregor Fitzi, and Jürgen Mackert

Urban Change and Citizenship in Times of Crisis
Volume 2: Urban Neo-liberalisation
Edited by Bryan S. Turner, Hannah Wolf, Gregor Fitzi, and Jürgen Mackert

Urban Change and Citizenship in Times of Crisis
Volume 3: Figurations of Conflict and Resistance
Edited by Bryan S. Turner, Hannah Wolf, Gregor Fitzi, and Jürgen Mackert

Urban Change and Citizenship in Times of Crisis

Volume 2: Urban Neo-Liberalisation

Edited by Bryan S. Turner, Hannah Wolf, Gregor Fitzi, and Jürgen Mackert

LONDON AND NEW YORK

First published 2020
by Routledge
2 Park Square, Milton Park, Abingdon, Oxon OX14 4RN

and by Routledge
52 Vanderbilt Avenue, New York, NY 10017

Routledge is an imprint of the Taylor & Francis Group, an informa business

© 2020 selection and editorial matter, Bryan S. Turner, Hannah Wolf, Gregor Fitzi and Jürgen Mackert; individual chapters, the contributors

The right of Bryan S. Turner, Hannah Wolf, Gregor Fitzi and Jürgen Mackert to be identified as the authors of the editorial material, and of the authors for their individual chapters, has been asserted in accordance with sections 77 and 78 of the Copyright, Designs and Patents Act 1988.

All rights reserved. No part of this book may be reprinted or reproduced or utilised in any form or by any electronic, mechanical, or other means, now known or hereafter invented, including photocopying and recording, or in any information storage or retrieval system, without permission in writing from the publishers.

Trademark notice: Product or corporate names may be trademarks or registered trademarks, and are used only for identification and explanation without intent to infringe.

British Library Cataloguing-in-Publication Data
A catalogue record for this book is available from the British Library

Library of Congress Cataloging-in-Publication Data
A catalog record has been requested for this book

ISBN: 978-0-367-20564-5 (hbk)
ISBN: 978-0-429-26228-9 (ebk)

Typeset in Times New Roman
by Wearset Ltd, Boldon, Tyne and Wear

Contents

List of figures ix
Notes on the contributors x

Introduction: urban warfare – neo-liberalism's assault on
democratic life in the city 1
HANNAH WOLF AND JÜRGEN MACKERT

PART I
Producing urban geographies of crisis 15

1 Revisiting territories of relegation: class, ethnicity, and state
  in the making of advanced marginality 17
  LOÏC WACQUANT

2 State-making as space-making: the three modes of the
  production of space in Istanbul 30
  SINAN TANKUT GÜLHAN

3 The city as a business 50
  NICOLE RUCHLAK AND CARSTEN LENZ

PART II
Governing cities in neo-liberalism 61

4 Restructuring Melbourne: uneven geographies of success 63
  SEAMUS O'HANLON

5 Governing through participation: activation of civil
   commitment in Berlin's neighbourhoods   78
   MAGDALENA OTTO

PART III
Everyday experience of urban neo-liberalisation   97

6 Permanent liminality? Housing insecurity and home   99
   HANNAH WOLF

7 Athens in times of crisis: experiences in the maelstrom of
   EU restructuring   119
   DINA VAIOU

8 The *right to the city* after Grenfell   131
   GARETH MILLINGTON

   Index   152

Figures

2.1	State-space mode I: world-imperial-quasi-colonial	36
2.2	State-space mode II: etatist/national developmentalist	38
2.3	State-space mode III: neo-liberal/state-corporate alliance	39
2.4	Construction permits issued by municipalities in Istanbul	40
2.5	State-led housing development projects in Istanbul	41
5.1	Schematic overlaps in the triangulation of societal spheres	80
5.2	Institutional realisation of participatory measures	82

Contributors

Sinan Tankut Gülhan is an Assistant Professor of Sociology at Gaziantep University. He has also chaired the graduate programme in urban studies since 2016. His current research interests include urban theory, urbanisation in Turkey, urban history, sociology of urbanisation, and critical perspectives in architectural history and urban planning.

Carsten Lenz has a PhD in Philosophy from the Ludwig Maximilians University in Munich. He works as an IT consultant. His main area of research is the sphere of economic-political discourse.

Jürgen Mackert is Professor of Sociology and a Co-Director of the Centre for Citizenship, Social Pluralism, and Religious Pluralism at the University of Potsdam, Germany. His research interests lie in the sociology of citizenship, political economy, closure theory, (neo-)liberalism, settler colonialism and de-democratisation. His most recent publication is *Populism and the Crisis of Democracy*, 3 vols., edited with Gregor Fitzi and Bryan S. Turner (2019).

Gareth Millington is a Senior Lecturer at the University of York. He recently published *Urban Criminology* (2019, Routledge) and *Urbanisation and the Migrant in British Cinema* (2016, Palgrave Macmillan). He has also published articles in the *British Journal of Sociology*, *The Sociological Review*, *City*, *International Journal of Urban and Regional Research*, *Antipode* and *Urban Studies*.

Seamus O'Hanlon is an Associate Professor at the School of Philosophical, Historical and International Studies (SOPHIS) at Monash University, Melbourne, where he teaches contemporary and urban history. He has published widely on urban history, economic history, and population culture in Australia and elsewhere in the twentieth century. His most recent book is *City Life: The New Urban Australia* (Sydney, 2018).

Magdalena Otto is a PhD student at the University of Potsdam. She holds a master's degree in the social sciences. Research interests: social inequality, integration, participation. Recent publication: S. Beigang, K. Fetz, D. Kalkum, and M. Otto. *Diskriminierungserfahrungen in Deutschland. Ergebnisse einer Repräsentativ- und einer Betroffenenbefragung*, Nomos, 2017.

Nicole Ruchlak holds a PhD in political science from the Ludwig Maximilians University in Munich. She works as a radio editor at the BR/German Public Radio. Her main area of research is the sphere of economic-political discourse.

Dina Vaiou is Professor Emeritus of Urban Analysis and Gender Studies in the Department of Urban and Regional Planning of the National Technical University of Athens. Her research interests include the feminist critique of urban analysis; the changing features of local labour markets, particularly women's work and informalisation processes; the impact of mass migration on Southern European cities.

Loïc Wacquant is Professor of Sociology and Research Associate at the Institute for Legal Research, Boalt Law School, University of California at Berkeley, where he is affiliated with the programme in medical anthropology, the Global Metropolitan Studies Program, and the Center for Urban Ethnography. He is also Researcher at the Centre de sociologie européenne in Paris.

Hannah Wolf is a researcher and lecturer completing her PhD at the University of Potsdam, coordinator of the 'Centre for Citizenship, Social Pluralism and Religious Diversity' and associate member at the collaborative DFG-research centre 'Re-Figuration of Spaces' (TU Berlin). Her research interests include political and moral economies of housing and home, citizenship, urban sociology, and the sociology of everyday life.

Introduction
Urban warfare – neo-liberalism's assault on democratic life in the city

Hannah Wolf and Jürgen Mackert

It comes as no surprise that for more than two decades now the global economic regime of neo-liberalism that operates as an economic theory, a political ideology, a policy paradigm, and a social imaginary (Evans & Sewell, 2013, p. 36) has had a serious impact on the city. Consequentially, sociological reflection has tried to conceptualise ongoing processes of transforming the city under the concept of *urban neo-liberalism* which is generally understood as the contextually specific and path-dependent realisation of neo-liberal restructuring projects, embedded in varying social, political, economic, and cultural 'regulatory landscapes' (Brenner & Theodore, 2002, p. 351). These discussions focus on how and to what effects neo-liberal strategies are employed in different spaces and contexts, as well as on their respective variegated counter-movements, since in this perspective cities are understood as both crucial sites of experimental implementation and of heightened resistance and protest. The processes of neo-liberalisation, coined as 'actually existing neo-liberalism' (Brenner & Theodore, 2002, p. 350) are by their very nature variegated and context-specific and can appear in multi-faceted and contradictory forms. Thus, we are not preoccupied with a kind of archetypical neo-liberal city but with complex and interwoven processes of urban neo-liberalisation yielding various forms of actually existing neo-liberal cities.

As debates on urban neo-liberalism have concentrated on processes of actual transformation of urban space, in the introduction to this volume we contend that the neo-liberal ideology at the heart of these processes must not be underestimated or all too easily discarded as merely an idea as opposed to actual empirical reality. Starting from a definition of neo-liberalism as a 'political project that is justified on philosophical grounds and seeks to extend competitive market forces, consolidate a market-friendly constitution and promote individual freedom' (Jessop, 2013, p. 70), we need to be concerned both with the philosophical grounds – the ontological assumptions of neo-liberalism as a belief system – and with the means by which the actors of the neo-liberal project pursue and aim to reach their goals – the strategies and techniques employed to marketise, competitise, and individualise social life. Additionally, we need to scrutinise the underlying logics and the all-encompassing effects of neo-liberalisation; and in this regard it might well be useful to take the words of but one representative of the neo-liberal global elite, Warren Buffett, very seriously: 'There's class warfare, all right, (...) but it's my class, the rich class, that's making war, and we're winning' (Stein, 2006). In this decades-old global war between the class of the rich and the class of the poor, neo-liberalism has tightened its grip on the city. In recent years, common efforts of global political and economic elites have made the city the lynchpin of intensified economic investment and put it centre stage in multi-faceted politico-economic strategies. Cities have been exposed to aggressive strategies of banks, hedge-funds, and investors and have been turned into the very terrain of profit-maximising strategies. Still, we have to look beyond the mere facts of speculation, land grabbing, and displacement to understand what this urban warfare is all about. In our understanding, the neo-liberal assault on cities around the globe is a war fought with different means or weapons and aiming at two main goals: *First*, and this is the obvious argument, the city itself is to turn into a space of investment in order to attract business, enterprises, and hedge-funds, which means that city administrations are to compete with one another vying for the favour of a highly potent global economic elite that is looking to make the most promising investment.

Second, and not so frankly mentioned in this war, neo-liberalism aims at fundamentally transforming human social order by forcing political citizens, the bearers of rights, to see themselves as no longer the true sovereigns of a democratic political community but simply as market subjects making choices – if they are financially capable of doing so. Neo-liberals are no longer ready to respect citizens' rights and entitlements to still existing public services but demand the final marketisation of a publicly financed infrastructure of health care and public education from kindergarten to university, services of energy, water, and waste disposal and, besides many other things and above all – the right to housing, the right to have a home as the basis of having and making a life in the city. Thus, urban neo-liberalism must be understood as the attempt to fully marketise and financialise city spaces by fundamentally de-democratising what is left of social life in the city. As both liberalism and neo-liberalism are deeply and constitutively anti-democratic ideologies (Macpherson, [1977] (2012); Biebricher, 2015; Brown, 2003, 2015) today's strategies of financial capital towards the city and its inhabitants – the citizens and consequently the democratic sovereign – are nothing but strategies of depriving these citizens of their rights.

We need to remind ourselves that neo-liberalism is not some obscure natural process happening due to a supposedly human nature, but in fact a political and economic strategy employed by specific actors and with specific aims. A look at the discourse of global elites after the 2008 global financial crisis can help to shed some light on this strategy: after the crisis that – perhaps ironically – started with a housing crisis initiated by investment bankers' deliberately created toxic financial instruments, the same global political and economic elites and billionaires that engulfed millions of people around the world in misery, have started their next attack upon citizens and democracy by cynically re-defining the city. Completely unimpressed by their collective failure, global financial elites – after their political lackeys had shifted the billion-dollar burden they had privately accumulated upon the shoulders of citizens – at the World Economic Forum (2016) announced a 'revolutionary' new comprehension of the city. Unsurprisingly, these self-declared 'Masters of the Universe' (Stedman Jones, 2012) proclaimed a city to be prospectively nothing but the exclusive site of guaranteed lucrative investment, rising returns, and maximum profits by speculation with space and living space and the very field of heightened *competition*.

This neo-liberal re-invention of the city poses a number of questions: first of all, what is the city, how and for whom is it imagined, planned, and produced? Second, by whom and by what institutional logics are cities governed in an age of neo-liberalism? And third, how do neo-liberalisation processes affect urban everyday life, how are they (re)produced, experienced and contested? Intimately linked with all these issues is, as we shall argue, the question of citizenship. Henri Lefebvre, whose 'right to the city' has recently regained prominence as the main slogan of many urban movements around the world, emphasised the need for an 'urban revolution' (1970) in order to counter a full economisation of space, social relations, and everyday life, and proposed the notion of *inhabitance* as the determining factor of citizenship: 'those who go about their daily routines in the city, both living in and creating urban space, are those who possess a legitimate right to the city' (Purcell, 2003, p. 577). In this understanding, urban citizenship is much more than paying your taxes and voting in local elections; it first of all rests on the recognition that the city is produced by those who inhabit it and who, by way of their inhabitance, have gained the right to collectively shape, appropriate and participate in the control over the use of urban space. Quite obviously, contemporary power structures suppress this legitimate right to the city and replace it by one based on financial, political and discursive power over the city-as-commodity and the citizen-as-consumer.

Imagining the city: commodification and competition

Neo-liberal ideas strongly contradict the modern city's nature, as the history of urbanisation shows. While the city of course has always been a space of economic activity and investment, in particular, it has been the critical space of democratic participation, the rise and dissemination of participatory ideas, of emancipatory class struggle and social movements as well as of critical conflicts in the face of social inequality, while migration made it the space of interethnic relations and cultural pluralism. All these processes have been part of the historical fight for citizenship, making the city the very place of the rise of democracy.

Thus, while industrialisation and capitalism alongside growing population density and intensified social interaction helped modern cities become not only motors of growth but also the focal point of political engagement, democratic demands and deliberation, today's transformation shows a totally new quality. Neo-liberalism's advocates' impoverished conceptions of both the city as a space for investment and competition and of the individual as a simple market subject poses a fierce attack on democratic civilisation in general and on urban life in particular: it aims at nothing less than suppressing, marginalising and finally destroying the manifold crossing social circles that for a long time have characterised urban life. It aims at putting an end to what the modern city has always been and still can be *apart* from economic activities.

This assault hardly comes as a surprise. As much as neo-liberalism as ideology and political programme aims at erasing any democratic participation in society, its proponents have taken sides pushing ahead the re-conceptualisation of the city as a market with the right of the stronger 'to do down the weaker' (Macpherson, [1977] 2012, p. 1). The visions of *what* a city ought to be as well as *how* and *for whom* it is to function are crucial in this respect. Neo-liberalism's understanding of competition as the basic mode of human and social existence deliberately distinguishes between grades of worthiness, of deserving and undeserving citizens and is thus opposed to any notion of unconditional rights. Neither a right to produce and transform the city, nor a right to accommodation and home, and no right to complain – this shows neo-liberalism's *anti-democratic* spirit becoming manifest in the city.

The consequences of the transformation of the city into a mainly economic space of market transactions have long since become visible in Global Cities like London, Paris, New York, Hong Kong, and many more: accommodation for the majority of citizens has become unaffordable, major parts of living space in the inner cities have been taken over by the wealthy, creating exclusive living spaces for them, or whole buildings in luxury districts are left vacant for speculation while social and political life has long since been defunct.

Yet, not only do some Global Cities lie at the heart of this economic strategy. Rather, in almost all urban agglomerations space, buildings, single flats and former industrial constructions have come into the focus of investors, as huge amounts of capital worldwide are looking for options of investment and return via a 'winner-take-all urbanism' (Florida, 2017, p. 13). Global economic elites, banks, and hedge-funds in times of low-interest-rate policy are pushing the transformation of cities into fields of strategic investment with high profit margins and declare those cities most capable of competing for potent investors, guaranteeing highest investment returns and offering state-backed guarantees of this property-to-speculate as *successful cities*. Success here is understood as the constant and 'future-proof' growth of urban economies and real estate markets and is measured e.g. by the JLL City Momentum Index that 'assesses a city's real estate market dynamics – its rates of construction and absorption, price movement, and the attractiveness of its built environment for cross-border capital sources and corporations' (World Economic Forum, 2016). *Successful cities* are part of a city-branding strategy employed by the World Economic Forum and business consultancy agencies such as McKinsey, PwC, and KPGM that have a very clear vision of the urban future consisting of *established, emerging*, and *new* world cities ready 'to compete successfully in a new and constantly evolving economic landscape' (World Economic Forum, 2016). Strikingly, questions of life *in* the city do not seem to play a role in this understanding of success. Yet, there are a number of other city brandings to be found – think of sustainable, resilient or smart cities – and political governments and their respective research institutions also play a role in shaping these terms and pursuing politico-economic agendas. Part of the *Urban agenda for the European Union*, for example, is a programme for smart cities, including an initiative to produce 'the smart citizen' (BBSR, 2017). Examples like this show that

> [cities] turn into a-historical spaces, where, for example, climate change takes place or mobility concepts, architectural styles, social projects or policies are implemented and applied. From this point of view, something is implemented into the city, and the city is reduced to a surface for applications and technologies of all kinds. The people living there are turned into recipients, users, residents or consumers.
> (Eckardt, 2017, p. 2., translation H. W. & J. M.)

Today, after citizens have got used to the consequences of austerity politics, which means tax-cuts for the wealthy but reduced wages, precarious employment, cut or loss of pensions, and reduction of public services for ordinary citizens (Blyth, 2013; Schäfer & Streeck, 2013) neo-liberals' attack on the city seems only consequential: in the ongoing global war of the rich class against the citizens, the global elite's strategy of *successful cities* is simply the urban dimension of this war. The global neo-liberal elite will certainly not declare the end of war until the city has been reshaped as a market place and the idea of citizens' sovereignty has vanished.

In a historical perspective, the rise and development of democracy and citizenship are most closely linked with a *political* understanding of the city as a space of democratic opinion formation, political deliberation, and consensus finding with regard to how a democratically governed citizenry wants to live and what citizens define as the common good. Therefore, the neo-liberal attack on and re-definition of the city as a space of naked market transactions and profit-maximising speculation not only threatens an urban lifeworld of citizens, who are either simply not included – as in the vision of *successful cities* – or forced to adapt their behaviour – as in the example of *smart citizens*; rather, it goes much deeper: what neo-liberal elites declare to be a *successful city* in fact is a strategy to de-politicise cities as crucial spaces of democratic life and to lever out citizens' participation and their rights of co-determination in social and democratic life. Nothing shall remain from the city as the centre of political contention and the primacy of political and democratic decisions on how we conceive of public and private space in the city.

A way to understand and conceptualise these processes is proposed by Loïc Wacquant in his contribution on to this volume. He develops the concept of urban relegation to describe and analyse the structural and often state-facilitated processes that serve to fundamentally marginalise and exclude citizens spatially, socially and symbolically. Urban relegation goes far beyond classical urban sociology's focus on spatial segregation into 'problematic' districts and ghettos – instead, it encompasses a set of institutionalised multi-scalar practices, relations and mechanisms that produce urban regimes of power and oppression.

Similarly poignantly, Sinan Tankut Gülhan offers a spatial framework to analyse the multi-faceted modes of neo-liberalised production of social space as state-space. For the case of Istanbul, he sheds important light on the role of state and government in the historical transformations shaping the city and its inhabitants' everyday life, and argues for an integrative perspective of *state spatiality* to understand both the material and symbolic restructuring of Istanbul.

One of the most startling examples is provided by Nicole Ruchlak and Carsten Lenz who discuss the imagination and actual creation of fully neo-liberalised cities-as-businesses. Based on the idea of market freedom as foremostly freedom from political control, economic elites are forming networks to build cities lacking any political or cultural history to consider and fully designed for extractive and profit-oriented purposes. Perhaps one of the most frightening insights lies in the fact that potential members of these 'cities' represent social/financial homogeneity justifying delusions of purity – no less radical and social Darwinist an idea than rising neo-Fascism that can be conceived as neo-liberalism's political/cultural counterpart. In their various disguises, these neo-liberal dreams of new cities for rich people as a-political spaces with regimes of supply and demand, design, and control, are ways to a dystopia and also a degradation of those who may have the financial means to afford to reside in Charter Cities, Free Private Cities, or seasteading projects.

Governing the city: uneven developments and institutional logics

If today the city has come under attack from neo-liberals, spaces of democratic deliberation and coalition building in favour of defending rights, as well as ideas of equality and individual freedom are in severe danger. As neo-liberals are convinced anti-democrats and thus declared opponents to democracy, democratic participation and democratic procedures, their grip on the city is a fierce and ruthless attack against democracy – for good reasons:

At its most rudimentary, modern democracy features universal equality and freedom. When democracy undergoes the economisation of state, society, and subject specific to contemporary neo-liberal rationality, these terms and practices are transmogrified. They lose their political valence and gain an economic one: freedom is reduced to the right to entrepreneurial ruthlessness and equality gives way to ubiquitously competitive worlds of winners and losers (Brown, 2016, p. 2).

This is what neo-liberalism is after: instead of conceiving the city as a place of social exchange, conflict and solidarity, political engagement, and democratic participation, the city as a market based on the pure principle of competition turns it into a de-politicised arena, while citizens' lives and political activities are replaced by competition:

> [The] economization of the political, and the reduction of citizenship to responsibilized investment in oneself, on the one hand, and to being human capital for the nation as firm, on the other, means that citizenship is stripped of substantive political engagement and voice, and citizen virtue becomes uncomplaining accommodation to the economic life of the nation.
>
> (Brown, 2016, p. 10)

Instead of trying to create an environment that makes the city a liveable place for all citizens and thus a goal of public politics, neo-liberals pledge to transform it into a battlefield of an *economically induced new war of all against all* – and, as Warren Buffet made clear, we already know the winners and the losers of this war. The logic behind putting people against one another on this battleground called the market place that they enter with extremely unevenly distributed weapons seems to be twofold again: first, as a choir of economists, politicians and the corporate media tell us 'there is no alternative' given the sovereign debt crises but to demand entrepreneurial activities from each and everybody under conditions of austerity; second, if people are kept busy with being permanently competitive in order to survive, they will not have any time to organise, to be politically active or make demands. Working conditions, contracts, pensions, and so forth can then be reduced without risking too much resistance. Increasing economic pressure is an easy means to de-politicise social life and to de-democratise the city and wider society.

In the end, what neo-liberals and their revolutionary idea of *successful cities* offer is a *dystopia* of life in the city devoid of any political or social agency; former citizens will have become subjugated to harsh exploitation, permanently competing with one another for accommodation, housing, jobs, health care, kindergarten, school, and so forth. Beyond this kind of degradation to the individual lives of citizens, the rat-race between cities as market places and spaces of investments has long since begun and the list of politicians and mayors, who made themselves the backers of capital in order to attract business, is indeed a very long one.

One very striking recent example of the urbanism co-created by business CEOs and politicians is Amazon's search for a location for their second headquarters, HQ2. Amazon's CEO Jeff Bezos had called for American cities to compete against each other in 'a contest to see who can give the most away' (Florida, 2017, p. 220). In this undignified contest, city leaders were outmatching each other at offering tax reductions to the giant tech corporation, among them New York Governor Andrew Cuomo who, after Amazon cancelled the plan to settle in New York – apparently due to public protests and the appointment of an HQ2 critic to the Public Authorities Control Board – wrote an open letter promising 'personal responsibility for the project's state approval' (New York Times, 2019) and is ready to do everything to have HQ2 established in New York City (Aronoff, 2019). Such strategies of currying favour with investors are neither without precedent nor beneficial for citizens – yet, they certainly are for business.

While incidents like these are widely publicised in the media, many decision-making processes tend to take place behind closed doors or are not as easily apparent because they happen over a far longer time-span. For the case of Melbourne, Seamus O'Hanlon reconstructs these long-term processes of uneven development as the result of a series of economic crises, the discursive power of civic and business leaders and the attempted re-creation of the city through imposing neo-liberal ideas of free market and post-industrial urban activities. His contribution shows how such urban 'revitalisation' projects tend to neglect local and communal resources and instead look for big solutions provided by big businesses, and by doing so in fact help to produce unequal urban geographies and distribution of life-chances within one metropolis.

A different aspect of institutional government is provided by Magdalena Otto: focusing on Berlin, she scrutinises the role of civil society and especially the narratives of citizens' participation and activation as part of a neo-liberalised policy agenda. Her contribution analyses the logics behind governing a Social City through participation and, through the lens of governmentality studies, how citizens' responsibilisation is achieved in an intermingling of top-down funding and financing with bottom-up activation and empowerment measures.

Everyday urban experience: the neo-liberal politics of asymmetrically distributed life-chances

What are citizens about to lose, or have already lost, as the city has come into the focus of neo-liberal strategies to turn it in a huge market place for investment? Nothing less, one might say, than the basic promises of modernity.

In a slightly modified quotation from Karl Marx' first sentence in *Capital I* (Marx, [1867] 1990), the wealth of Western societies appears as an immense accumulation of *life-chances*. This simple statement reminds us that modernity started not only with the promise of abolishing inequalities and offering options to enable people to make a living but also with the promise of endowing equal rights to every citizen. Liberal democracy has been proven far from perfect or even a satisfying model in this regard. Yet, there were options to further develop this structural and cultural concept and to broaden citizens' opportunities (Marshall, 1950). Far from being a 'socialist' system, as neo-liberals are happy to discredit liberal democratic systems with developed institutions of a welfare state, everyday political, social, cultural, and economic life was far from paradise – simply, as Max Weber stressed, 'because conflict [*Kampf* (!) in German original, translation H. W. & J. M.] is an ineradicable element of all cultural life' (Weber, 2012, p. 320).

At first sight this might seem in accordance with neo-liberalism's Social Darwinism, yet, in a sociological perspective, it is obvious that Weber's 'cultural life' refers to different 'spheres of values' that are irreducible to a single one, contrary to what neo-liberals argue. This is not only an obvious logical inconsistency but nothing less than a political strategy to deny the political, social, and cultural dimensions of the struggle for life-chances. The theoretical poverty of neo-liberalism becomes obvious here, and its strategy to degrade the plurality and diversity of all dimensions of social life in order to subsume it to an allegedly disdainful neutral market mechanism could be called a simple absurdity.

While liberal democracy in the *trente glorieuse* of Western societies was a system that at least in principle opened different fields to citizens' struggles for life-chances, today neo-liberalism makes no secret of the fact that it is not interested at all in opening up chances for each and everybody. Thus, while already under conditions of liberal democracy there has always been a huge gap between the privileged and those who are worse off, as individuals are located in extremely different positions in the social structure equipped with different resources or capital to start their struggle for chances in life (Bourdieu, 1984; Bourdieu & Passeron, 1977), neo-liberal politics have aggravated this situation by implementing and pursuing a radical politics of asymmetrically distributed life-chances. The dynamics of these politics are well known: tax-cuts for the wealthy and a huge project of redistribution of wealth in societies to the disadvantage of both the middle classes and the poor, the consequence being growing inequality, poverty, exclusion and conflicts, and finally the denial of ordinary citizens' 'right to have rights' and to make a living in Western capitalist societies.

In this sense, neo-liberalism is a politics of extreme social exclusion, and here as well it meets with populist and neo-Fascist politics. While the latter pursue a politics of the radical exclusion of the cultural and political 'other', neo-liberalism follows the idea of the expulsion or social cleansing of those who cannot keep up in a world of pure economic and financial Social Darwinism. Ideologies of self-responsibility and presumed equality of opportunities divide people along lines of deservingness, and they blame and shame those who struggle to make a living at the bottom rungs of society. What urban neo-liberal strategists are therefore looking for is to strictly separate the economic elite from the rest of the population, to appropriate and occupy the best places and positions, to create gated communities, have them protected by private security firms, and reserve resources for themselves.

This of course is bad news for citizens on this globe. It is hardly necessary to remind anybody how many people all over the world today live in misery because of neo-liberalism and the criminal greed of its sorcerer's apprentices in banks, global investment and hedge-funds. Millions of citizens are without stable housing and dwelling in degrading living conditions because of a financial elite that is interested in nothing but senseless speculation in financial markets. Basic human needs such as housing and health have been turned into 'fictitious commodities' (Polanyi, [1944] 2001) and are viewed as assets ready for capital extraction. As the global financial elite has been so successful with putting at stake the lives of so many millions of people and families, why not take this momentum and get rid of the superfluous people in the city by tightening the neo-liberal grip on the houses and homes of the miserable? It is this simple logic that lies behind neo-liberalism's attack on the city – driving people out of their houses and flats, raising property prices and rents that no one can afford, and turning the city into a space of investment. The superfluous may then enter into the war of all against all caused by an unjustifiable and criminal economic system and compete for the crumbs left over.

What is left from modernity's promise of more life-chances for everybody, less social inequality and full political equality? Not too much, we have to concede; rather, we see the contrary as a kind of neo-Feudal system (Neckel, 2010) has been established where neo-liberalism's ideology and arbitrary power has taken the place of the Church in pre-modern times as a second power centre beside the state (with the small qualification that neo-liberal governance today is much more characterised by a smoothly organised cooperation of state and economy than the fierce competition of the state and Church in former times). While modernity made a plea for equality, neo-liberalism is in favour of extreme inequality, it protects both inherited wealth and privilege, and it has succeeded in again conflating the disposal of financial means with political decision-making. In neo-liberal 'society' the wealthy are making decisions in favour of the wealthy; only the happy few shall have rights to act politically, while the rest have to try to make ends meet. And, like in Feudalism, ordinary people simply do not play any role. Yet, there is one big difference: while neo-liberalism takes all the advantages of a neo-Feudal system, it rejects any kind of moral regulation of the economy as it has always been the case in pre-modern times.

Today, we might be back in a time of fighting with whatever is needed to defend a democratic civilisation and face the power of the global neo-liberal elite. Their politics of asymmetrically distributed life-chances aims at nothing less than fundamentally restricting citizens' life-chances, their political voice and participation, and thus putting an end to democracy itself. However, the distinction between those actors who pursue and push forward this project and those who resist and protest against it might not be drawn very easily. Neo-liberalisation is a process that takes place at different scales and in different spheres of social life and is indeed deeply embedded in everyday life. The capability to not be a part of this project might in the end be an illusion given the extent of individual responsibilisation and competition that citizens face and exert in their lives.

In her contribution, Hannah Wolf deals with citizens' experiences of neo-liberal attacks on arguably one of the most fundamental human needs: housing and home. For the cases of London and Berlin she reconstructs the long-term decisions and processes that have produced a severe housing crisis and shows how the interplay of politico-economic policies and ideologies along with imperatives and practices of self-responsibility have turned the foundation of human and social existence into a state of permanent liminality in which not only people's right to housing but also their capability to feel at home is fundamentally at stake. Reconsidering the concept of crisis itself, her chapter draws attention to the complex experiences of insecurity that 'The Housing Crisis' produces.

With a focus on post-crisis austerity politics in Greece, Dina Vaiou discusses the effects of austerity politics on the everyday lives and experiences of women in Athens. She argues how vital it is to hear and give recognition to the voices of the marginalised and to understand the crisis through the life-stories of those affected. Not only does she show the progressive restriction and denial of life-chances affecting important dimension of social life – labour, housing, and care – she also discusses women's agencies and subjectivities under conditions of enforced austerity and struggle.

Finally, Gareth Millington brings up one of the most shocking and destructive examples of urban neo-liberalism, the catastrophe of London's Grenfell Tower, a block of social housing situated in the wealthiest part of London. Not only does he show how political neglect to hear the many complaints and warnings by the residents and the unwillingness to provide standard safety materials for social housing tenants has led to the death of nearly 80 individuals, he also discusses how a Lefebvrian notion of the 'right to the city' needs to incorporate issues of protection and safety instead of sticking to its inherent scepticism towards the state and bureaucracy in order to realise the said right for low-income citizens.

After neo-liberalism: the Social City?

The city has become a focal point for neo-liberalism's war against democracy and citizens. What we see in cities today is neo-liberalism's most hideous face, subordinating everything to the principle of profit-making, as competition becomes the only legitimate way to organise social life. In this perspective, it is obvious that durable social inequalities will keep on growing and get inscribed into politics; trust in democratic procedures will vanish while financial capital will further withdraw from any kind of regulation by states and public politics will come to an end as it has long since given up any idea of a politics that is in the favour and good of all citizens, and not only for corporate business and investment. In this situation societies de-democratise, as we see today in the face of growing neo-fascism in many cities, as competition dominates daily life, life-chances are getting more and more asymmetric, and the idea of the 'social' is once again under attack from neo-liberal elites.

Turning *social* relations into market transactions in order to restructure today's cities is not a new idea from the neo-liberals but one of the non-negotiable dogmas of their religion called science. For neo-liberalism any idea of a 'Social City' must be a pure nightmare. If we look back today at the destruction of social security and safety nets in Western societies in the 1980s, the early fights of Conservatives in the UK (King & Waldron, 1988), the spearheads of a radical attack on citizenship rights, against entitlements and the welfare state, were not just a critique of welfare institutions that allegedly had grown exuberantly but a fundamental attack against everything that we conceive to be *social* and that neo-liberalism aims to transform into pure market transactions. Neo-liberalism's chief ideologue, Friedrich von Hayek, rejected almost any form of social protection that was intended to support and enable people to participate in society or to promote a democratic society in which *every* citizen might be able to make a living. In order to establish the market as the only legitimate principle of organising any kind of social relation, Hayek conceived state involvement to regulate economic activities as an illegitimate force or constraint. In order to show his disgust he despised the word 'social' as a so-called 'weasel word'. Once added to other words it would deprive them – as the weasel does with eggs – of both their content and sense:

> It is true, that a social market economy is no market economy, a social rule of Law is no rule of Law, a social conscience is no conscience, social justice is not justice – and, I am afraid, a social democracy also is no democracy.
> (Hayek, [1979] 2004, pp. 61–62; translated H. W. & J. M.)

The 'godfather' of neo-liberalism would fiercely discredit the idea of citizens' rights to have homes, houses or flats for reasonable rents or prices, support in difficult situations, social security or – today – even a basic income in order to make a decent living, as pure socialist totalitarianism, degrading these people, enslaving them, and sending them onto the 'road to serfdom' (Hayek, [1944] 2007). It comes as no surprise that Margaret Thatcher – Hayek's most diligent political disciple – in the 1980s implemented this conviction into her political programme as UK Prime Minister:

> I think we have been through a period when too many people have been given to understand that when they have a problem it is government's job to cope with it. 'I have a problem, I'll get a grant. I'm homeless, the government must house me'. They are casting their problems on society. And, you know, *there is no such thing as society*. There are individual men and women and there are families. And no governments can do anything except through people, and people must look to themselves first. It is our duty to look after ourselves and then, also, to look after our neighbours. People have got their entitlements too much in mind, without the obligations. There is no such thing as an entitlement, unless someone has first met an obligation.
> (Thatcher, 1987, emphasis added)

Against the proclaimed doctrine of deregulation, however, so much of the literature on neo-liberalism (e.g. Brenner, Peck, & Theodore, 2010; Jessop, 2013) and the contributions in this volume show that the supposed freedom from state regulation is in fact a *re-regulation* of political means in favour of an economisation and marketisation of society. This insight helps to understand neo-liberalism as a political project – and it can and has to be the role and responsibility of politics to regulate in a different direction. To achieve this, it is necessary to keep in mind that despite the place- and context-specific individual realisations and processes of neo-liberalisation, citizens all over the world are not fighting against single strategies but against neo-liberalism as a fundamentalist religion. As much as the global neo-liberal elite of investment bankers, hedge-fund managers, billionaires, and politicians have been building coalitions-in-competition, citizens need to build supra-local alliances in solidarity against this destruction of the social fabric of cities and societies, against the attempts to de-politicise and de-democratise them and against the urban warfare against the poor that once were citizens.

References

Aronoff, K. (2019). If democrats lure back Amazon to NYC, they won't be forgiven. *New York Times*. Tuesday 5 March, 2019. Retrieved from www.theguardian.com/commentisfree/2019/mar/05/if-democrats-lure-amazon-back-to-nyc-they-wont-be-forgiven.

BBSR (Federal Institute for Research on Building, Urban Affairs and Spatial Development) (2017). On the road to becoming a smart citizen. BBSR Analysen Kompakt 08/2017. Retrieved from https://ec.europa.eu/futurium/sites/futurium/files/on_the_road_to_becoming_a_smart_citizen.pdf.

Biebricher, T. (2015). Neoliberalism and democracy. *Constellations*, 22(2), 255–266. doi: 10.1111/1467-8675.12157.

Blyth, M. (2013). *Austerity: The history of a dangerous idea*. Oxford: Oxford University Press.

Bourdieu, P. (1984). *Distinction. A social critique of the judgement of taste*. London: Routledge, Kegan & Paul.

Bourdieu, P., & Passeron, J.-C. (1977). *Reproduction in education, society and culture*. London, Thousand Oaks, CA, and New Delhi: Sage.

Brenner, N., & Theodore, N. (2002). Cities and the geographies of 'actually existing neoliberalism'. *Antipode*, 34, 349–379.

Brenner, N., Peck, J., & Theodore, N. (2010). After neoliberalization? *Globalizations*, 7(3), 327–345.

Brown, W. (2003). Neo-liberalism and the end of liberal democracy, *Theory and Event*, 7. Retrieved from https://muse.jhu.edu/article/48659.

Brown, W. (2015). *Undoing the demos: Neoliberalism's stealth revolution*, Cambridge and London: Zone Books.

Brown, W. (2016). Sacrificial citizenship: Neoliberalism, human capital and austerity politics. *Constellations*, 23(1), 2–14.

Eckardt, F. (2017). *Schlüsselwerke der Stadtforschung*. Wiesbaden: Springer.

Evans, P. B., & Sewell Jr., W. H. (2013). Neo-liberalism: Policy regimes, international regimes, and social effects. In P. A. Hall & M. Lamont (Eds.), *Social resilience in the neo-liberal era* (pp. 35–68). Cambridge: Cambridge University Press.

Florida, R. (2017). *The new urban crisis: How our cities are increasing inequality, deepening segregation, and failing the middle class – and what we can do about it*. New York, NY: Basic Books.

Hayek v., F. A. [1979] (2004). Wissenschaft und Sozialismus. In F. A. v. Hayek, *Gesammelte Schriften in deutscher Sprache, Bd. 7* (pp. 52–62). Tübingen: Mohr Siebeck.

Hayek v., F. A. [1944] (2007). *The road to serfdom*. Chicago, IL: The University of Chicago Press.

Jessop, B. (2013). Putting neoliberalism in its time and place: A response to the debate. *Social Anthropology*, 21(1), 65–74.

King, D. S., & Waldron, J. (1988). Citizenship, social citizenship and the defence of welfare provision. *British Journal of Political Science*, 18(4), 415–443.

Lefebvre, H. (1970). *La révolution urbaine*. Paris: Gallimard.

Macpherson, C. B. [1977] (2012). *The life and times of liberal democracy*, Oxford: Oxford University Press.
Marshall, T. H. (1950). *Citizenship and social class and other essays*. Cambridge: Cambridge University Press.
Marx, K. [1867] (1990). *Capital. Volume I*. London: Penguin Books.
Neckel, S. (2010). Refeudalisierung der Ökonomie: Zum Strukturwandel kapitalistischer Wirtschaft. *MPIfG Working Paper 10/6*. Retrieved from www.mpifg.de/pu/workpap/wp10-6.pdf.
Polanyi, K. [1944] (2001). *The great transformation: The political and economic origins of our time*. Boston, MA: Beacon Press.
Purcell, M. (2003). Citizenship and the right to the global city: Reimagining the capitalist world order. *International Journal of Urban and Regional Research, 27*(3), 564–590.
Schäfer, A., & Streeck, W. (Eds.) (2013). *Politics in the age of austerity*. Cambridge: Polity Press.
Stedman Jones, D. (2012). *Masters of the universe: Hayek, Friedman, and the birth of neoliberal politics*. Princeton, NJ: Princeton University Press.
Stein, B. (2006). *In class warfare guess which class is winning*. Retrieved from www.nytimes.com/2006/11/26/business/yourmoney/26every.html.
Thatcher, M. (1987). *Interview for 'Women's Own'*. Retrieved under www.margaretthatcher.org/document/106689.
Weber, M. (2012). *Collected methodological writings*. H. H. Bruun & S. Whimster (Eds). London and New York, NY: Routledge.
World Economic Forum (2016). *Which are the world's 20 most successful cities?* Retrieved from www.weforum.org/agenda/2016/01/what-does-it-mean-to-be-a-successful-city/.

Part I
Producing urban geographies of crisis

1 Revisiting territories of relegation
Class, ethnicity, and state in the making of advanced marginality

Loïc Wacquant

Introduction

To relegate (from the late Middle English, *relegaten*, meaning to send away, to banish) is to assign an individual, population or category to an obscure or inferior position, condition or location.[1] In the post-industrial city, relegation takes the form of real or imaginary consignment to distinctive socio-spatial formations variously and vaguely referred to as 'inner cities', 'ghettos', 'enclaves', 'no-go areas', 'problem districts', or simply 'rough neighbourhoods'. How are we to characterise and differentiate these spaces, what determines their trajectory (birth, growth, decay, and death), whence comes the intense symbolic taint attached to them at century's edge and what constellations of class, ethnicity, and state do they both materialise and signify? These are the questions I pursued in my book *Urban outcasts* through a methodical comparison of the trajectories of the black American ghetto and the European working-class peripheries in the era of neo-liberal ascendancy (Wacquant, 2008a).[2] In this chapter, I revisit this cross-continental sociology of 'advanced marginality' to tease out its broader lessons for our understanding of the tangled nexus of symbolic, social, and physical space in the polarising metropolis at the century's threshold in particular, and for comparative urban studies in general.

To speak of *urban relegation* – rather than 'territories of poverty' or 'low-income community', for instance – is to insist that the proper object of inquiry is not the place itself and its residents but the multilevel structural processes whereby persons are selected, thrust, and maintained in marginal locations, as well as the social webs and cultural forms they subsequently develop therein. Relegation is a *collective activity*, not an individual state; a *relation* (of economic, social, and symbolic power) between collectives, not a gradational attribute of persons. It reminds us that, to avoid falling into the false realism of the ordinary and scholarly common sense of the moment, the sociology of marginality must fasten not on vulnerable 'groups' (which often exist merely on paper, if that) but on the *institutional mechanisms* that produce, reproduce, and transform the network of positions to which its supposed members are dispatched and attached. And it urges us to remain agnostic as to the particular social and spatial configuration assumed by the resulting district of dispossession. In particular, we cannot presume that the emerging social entity is a 'community' (implying at minimum a shared surround and identity, horizontal social, and common interests), even a community of fate, given the diversity of social trajectories that lead into and out of such areas.[3] We should also not presuppose that income level or material deprivation is the preeminent principle of vision and division, as persons with low income in any society are remarkably heterogeneous (artists and the elderly, service workers and graduate students, the native homeless and paperless migrants, etc.) and form at best a statistical category.

Urban outcasts is the summation of a decade of theoretical and empirical research tracking the causes, forms, and consequences of urban 'polarisation from below' in the United States and Western Europe after the close of the Fordist–Keynesian era, leading to a diagnosis of the predicament of the *post-industrial precariat* coalescing in the neighbourhoods of relegation of advanced society. The book brings the core tenets of Bourdieu's sociology to bear on a wide array of field, survey, and historical data on inner Chicago and outer Paris to contrast the sudden implosion of the black American ghetto after the riots of the 1960s with the slow decomposition of the working-class districts of the French urban periphery in the age of deindustrialisation. It puts forth three main theses and sketches an analytic framework for renewing the comparative study of urban marginality that I spotlight to help us elucidate the relations of poverty, territory, and power in the post-industrial city.

From ghetto to hyperghetto, or the political roots of black marginality

The study opens by parsing the reconfiguration of race, class, and space in the American metropolis because the foreboding figure of the dark ghetto has become epicentral to the social and scientific imaginary of urban transformation at the turn of the century.[4] On American shores, the abrupt and unforeseen involution of the 'inner city' – a geographic euphemism obfuscating the reality of the ghetto as an instrument of ethno-racial entrapment imposed uniquely upon blacks – was the target of a fresh plank of policy worry and scholarly controversy. Across Western Europe, vague images of 'the ghetto' as a pathological space of segregation, dereliction, and deviance imported from America (with rekindled intensity after the Los Angeles riots of Spring 1992) suffused as well as obscured journalistic, political, and intellectual debates on immigration and inequality in the dualising city.

The first thesis, accordingly, charts the *historical transition from ghetto to hyperghetto* in the United States and stresses the pivotal role of state structure and policy in the (re)production of racialised marginality. Revoking the trope of 'disorganisation' inherited from the Chicago school of the 1930s and rejecting the tale of the 'underclass' (in its structural, behavioural, and neo-ecological variants) which had come to dominate research on race and poverty by the 1980s, *Urban outcasts* shows that the black American ghetto collapsed after the peak of the civil rights movement to spawn a novel organisational constellation: the hyperghetto. To be more precise, the 'Black Metropolis', lodged in the heart of the white city but cloistered from it, which both ensnared and enjoined African-American urbanites in a reserved perimeter and a web of shared institutions built by and for blacks between 1915 and 1965,[5] collapsed to give way to a *dual socio-spatial formation*.

This decentred formation, stretching across the city, is composed of the *hyperghetto* proper (HyGh); that is, the vestiges of the historic ghetto now encasing the precarised fractions of the black working-class in a barren territory of dread and dissolution devoid of economic function and doubly segregated by race and class, on the one hand, and of the burgeoning *black middle-class districts* (BMCD) that grew mostly via public employment in satellite areas left vacant by the mass exodus of whites to the suburbs, on the other. Whereas space unified African Americans into a compact if stratified community from World War I to the revolts of the 1960s, now it fractures them along class lines patrolled by state agencies of social control increasingly staffed by middle-class blacks charged with overseeing their unruly lower-class brethren.[6] The encapsulating dualism of the Fordist half-century inscribed in symbolic, social, and physical space, summed up by the equation White:Black :: City:Ghetto has thus been superseded by a more complex and tension-ridden structure White:Black :: City::BMCD:HyGh following a fractal logic according to which the residents of the hyperghetto find themselves doubly dominated and marginalised.

Breaking with the stateless cast of mainstream US sociology of race and poverty, *Urban outcasts* then finds that hyper-ghettoisation is economically underdetermined and politically overdetermined. The most distinctive cause of the extraordinary social intensity and spatial concentration of black dispossession in the hyperghetto is not the 'disappearance of work' (as argued by William Julius Wilson) or the stubborn persistence of 'hyper-segregation' (as proposed by Douglas Massey), although these two forces are evidently at play (Wilson, 1996; Massey & Denton, 1993). It is government *policies of urban abandonment* pursued across the gamut of employment, welfare, education, housing, and health at multiple scales, federal, state, and local, and the correlative breakdown of public institutions in the urban core that has accompanied the downfall of the communal ghetto. This means that the conundrum of class and race (as denegated ethnicity) in the American metropolis cannot be resolved without bringing into our analytic purview the shape and operation of the state, construed as a stratification and classification agency that decisively shapes the life options and strategies of the urban poor.

The 'convergence thesis' specified and refuted

The second part – and central thesis – of *Urban outcasts* takes the reader across the Atlantic to disentangle the same spatial nexus of class, ethnicity, and state in post-industrial Europe. Puncturing the panic discourse of 'ghettoisation' that has swept across the continent over the past two decades, crashing into the Nordic countries head-on in the 2000s,[7] it demonstrates that zones of urban deprivation in France and neighbouring countries are not ghettos *à l'américaine*. Despite surface similarities in social morphology (population makeup, age mix, family composition, relative unemployment, and poverty levels) and representations (the sense of indignity, confinement, and blemish felt by their residents) due to their common position at the bottom of the material and symbolic hierarchy of places that make up the metropolis, the remnants of the black American ghetto and European working-class peripheries are separated by enduring differences of structure, function, and scale as well as by the divergent political treatments they receive.

To sum them up: repulsion into the black ghetto is determined by ethnicity (E), inflected by class (C) with the emergence of the hyperghetto in the 1970s, and intensified by the state (S) throughout the century, according to the algebraic formula $[(E>C) \times S]$. By contrast, relegation in the urban periphery of Western Europe is driven by class position, inflected by ethnonational membership and mitigated by state structures and policies, as summed up by the formula $[(C>E) \div S]$. It is not spawning 'immigrant cities within the city', endowed with their own extended division of labour and duplicative institutions, based on ethnic compulsion applied uniformly across class levels. It is not, in other words, converging with the black American ghetto of the mid-twentieth century characterised by its joint function of social ostracisation and economic exploitation of a dishonoured population.

To lump variegated spaces of dispossession in the city under the label of 'ghetto' bespeaks, and in turn perpetuates, three mistakes that the book dispels. The first consists in invoking the term as a mere rhetorical device intended to shock public conscience by activating the lay imaginary of urban badlands.[8] But a ghetto is not a 'bad neighbourhood', a zone of social disintegration defined (singly or in combination) by segregation, deprivation, dilapidated housing, failing institutions and the prevalence of vice and violence. It is a *spatial implement of ethno-racial closure and control* resulting from the reciprocal assignation of a stigmatised category to a reserved territory that paradoxically offers the tainted population a structural harbour fostering self-organisation and collective protection against brute domination (for elaborations, see Wacquant, 2008b, 2011). The second mistake consists in conflating the communal ghetto with the hyperghetto: impoverishment, economic informalisation, institutional desertification and the de-pacification of everyday life are not features of the ghetto but, on the contrary, *symptoms of its disrepair and dismemberment*.

The third error misreads the evolution of traditional working-class territories in the European city. In their phase of post-industrial decline, these defamed districts have grown more ethnically heterogeneous while postcolonial migrants have become more dispersed (even as nodes of high density have emerged to fixate media attention and political worry) (Pan Ké Shon & Wacquant, 2012);[9] their organisational ecology has become more sparse, not more dense; their boundaries are porous and routinely crossed by residents who climb up the class structure; and they have failed to generate a collective identity for their inhabitants – notwithstanding the fantastical fear, coursing through

Europe, that Islam would supply a shared language to unify urban outcasts of foreign origins and fuel a process of 'inverted assimilation'. (Liogier, 2012). In each of these five dimensions, neighbourhoods of relegation in the European metropolis are consistently *moving away from the pattern of the ghetto* as a device for socio-spatial enclosure: they are, if one insists on retaining that spatial idiom, *anti-ghettos*.

To assert that lower-class districts harbouring high densities of bleak public housing, vulnerable households, and postcolonial migrants are not ghettos is not to deny the role of ethnic identity – or assignation – in the patterning of inequality in contemporary Europe. *Urban outcasts* is forthright in stressing the 'banalization of venomous expressions of xenophobic enmity' and the 'cruel reality of durable exclusion from and abiding discrimination on the labour market' based on national origins; it fully acknowledges that 'ethnicity has become more a more salient marker in French social life' (pp. 195–196) as in much of the continent. But *cognitive salience is not social causation*. The sharp appreciation of the ethnic currency in the political and journalistic fields does not mean that its weight has grown *pari passu* as a determinant of position and trajectory in the social and urban structure, nor that it now routinely skews ordinary interactions and everyday experience.[10] Moreover, ethnic rifts, when they do surge and stamp social relations, do not assume everywhere the same material form.

To maintain that ghettoisation is *not* at work in the pauperised and stigmatised districts of the European city is simply to recognise that the modalities of ethno-racial classification and stratification, including their inscription in space, differ on the two sides of the Atlantic, in keeping with long-standing differences in state, citizenship, and urbanism between Western Europe and the United States. On the urban periphery of the Old World, resurging or emerging divisions based on symbolic markers activated by migration do not produce 'ethnic communities' in the Weberian sense of segmented collectives, ecologically separate and culturally unified, liable to act as such on the political stage,[11] as the inflexible hypodescent-based cleavage called race does for African Americans – and only for them in the sweep of history in the country. Ethnicity is defined by shifting and woolly criteria that operate inconsistently across institutional domains and levels of the class structure, such that it does not produce a coordinated alignment of boundaries in symbolic, social, and physical space liable to foster a dynamic of ghettoisation.[12]

The 'emergence thesis' formulated and validated

Refuting the thesis of transatlantic convergence on the pattern of the black American ghetto leads to articulating the thesis of the *emergence of a new regime of urban marginality*, distinct from that which prevailed during the century of industrial growth and consolidation running roughly from 1880 to 1980. The third part of *Urban outcasts* develops an ideal-typical characterisation of this ascending form of 'advanced marginality' – thus called because it is not residual, cyclical or transitional, but rooted in the deep structure of financialised capitalism – that has supplanted both the dark ghetto in the United States and traditional workers' territories in Western Europe.[13] A cross-sectional cut reveals six synchronic features (Chapter 8) while a longitudinal perspective ferrets out four propitiating dynamics (Chapter 9), including the polarisation of the occupational structure and the reengineering of the state to foster commodification. Here I want to spotlight two of those features, the one material and the other symbolic, to emphasise the novelty of advanced marginality.

The paramount material attribute of the emerging regime of marginality in the city is that it is *fed by the fragmentation of wage labour*, that is, the diffusion of unstable, part-time, short-term, low-pay, and dead-end employment at the bottom of the occupational structure – a master trend that has accelerated and solidified across advanced nations over the past two decades (Cingolani, 2011; Kalleberg, 2011; Pelizzari, 2009). Whereas the life course and household strategies of the working-class for much of the twentieth century were anchored in steady industrial employment set by the formula 40–50–60 (40 hours a week for 50 weeks of the year until age 60, in rough international averages), today the unskilled fractions of the deregulated service proletariat face a simultaneous dearth of jobs and plethora of work tenures that splinter and destabilise them. Their temporal horizon is shortened as their social horizon is occluded by the twin obstacles of endemic unemployment and rampant precarity, translating into the conjoint festering of hardship and proliferation of the 'working poor'.[14]

This double economic penalty is particularly prevalent in lower-class neighbourhoods gutted out by deindustrialisation. One illustration: in France between 1992 and 2007, the number of wage earners in insecure jobs (short-term contracts, temporary slots, government-sponsored posts, and traineeships) increased from 1.7 million to 2.8 million to reach 12.4 per cent of the active workforce against the backdrop of a national unemployment rate oscillating between 7 and 10 per cent; for those ages 15 to 24, that proportion jumped from 17 to 49 per cent (Maurin & Savidan, 2008). But, in the 571 officially designated 'sensitive urban zones' (ZUS) targeted by France's urban policy, the combined share of unemployed and precariously employed youths zoomed from 40 per cent in 1990 to above 60 per cent after 2000. Far from protecting from poverty as it expands, fragmented wage labour is a vector of *objective* social insecurity among the post-industrial proletariat as well as *subjective* social insecurity among the inferior strata of the middle-class – whose members fear social downfall and proving unable to transmit their status to their children due to intensified school competition and the loosening of the links between credentials, employment, and income. On this count, *Urban outcasts* is an invitation to *relink class structure and urban structure* from the ground up and a warning that an exclusive focus on the spatial dimension of poverty (as fostered, for instance, by studies of 'neighbourhood effects')[15] partakes of the obfuscation of the new social question of the early twenty-first century: namely, the spread and normalisation of social insecurity at the bottom of the class ladder and its ramifying impact on the life strategies and territories of the urban precariat.

But the inexorable propagation of 'McJobs' – *petits boulots* in France, *Billig-Jobs* in Germany, 'zero-hour contracts' in the United Kingdom, *lavoretti* in Italy, *biscate* in Portugal, etc. – is not the only force impinging on the precariat. A second, properly symbolic vector acts to entrench the social instability and redouble the cultural liminality of its constituents: *territorial stigmatisation*. Mating Bourdieu's theory of symbolic power with Goffman's analysis of the management of spoiled identities (Bourdieu, 1990; Goffman, 1964), I forged this notion to capture how the blemish of place affixed on zones of urban decline at the turn of the century affects the sense of self and the conduct of their residents, the actions of private concerns and public bureaucracies, and the policies of the state toward dispossessed populations and districts in advanced society. First, I document that the territorial taint is indeed a distinctive, novel, and generalised phenomenon, correlative of the dissolution of the black American ghetto and of the European working-class periphery of the Fordist-Keynesian period, that has become superimposed on the stigmata traditionally associated with poverty, lowly ethnic origins, and visible deviance. Since the publication of the book, proliferating studies have documented the rise, tenacity, and ramifying reverberations of spatial stigma in cities spread across three continents.[16]

Next, I show that the denigration of place wields causal effects in the dynamics of marginality via cognitive mechanisms operating at multiple levels. Inside districts of relegation, it incites residents to engage in coping strategies of mutual distancing, lateral denigration, retreat into the private sphere and neighbourhood flight that converge to foster diffidence and disidentification, distend local social ties, and thus curtail their capacity for proximate social control and collective action. Around them, spatial disgrace warps the perception and behaviour of operators in the civic arena and the economy (as when firms discriminate based on location for investment and address of residence for hiring),[17] as well as the delivery of core public services such as welfare, health, and policing (law-enforcement officers feel warranted to treat inhabitants of lowly districts in discourteous and brutal manner). In the higher reaches of social space, territorial stigma colours the output of specialists in cultural production such as journalists and academics; it contaminates the views of state elites, and through them the gamut of public policies that determine marginality upstream and distribute its burdens downstream. To label a depressed cluster of public housing a '*cité-ghetto*', a 'sink estate', or a '*ghetto-område*' fated by its very makeup to devolve into an urban purgatory closes off alternative diagnoses and facilitates the implementation of policies of removal, dispersal or punitive containment.[18]

Lastly, I propose that territorial stigmatisation actively contributes to *class dissolution* in the lower regions of social and physical space. The sulphurous representations that surround and suffuse declining districts of dispossession in the dual metropolis reinforce the objective fragmentation of the post-industrial proletariat stemming from the combined press of employment precarity, the shift from categorical welfare to contractual workfare and the universalisation of secondary schooling as a path to access even unskilled jobs. Spatial stigma robs residents of the ability to claim a place and fashion an idiom of their own; it saddles them with a noxious identity, imposed from the outside, which adds to their symbolic pulverisation and electoral de-valourisation in a political field recentred around the educated middle-class. So much to say that the precariat is *not* a 'new dangerous class', as proposed by Guy Standing (2011), but a miscarried collective that can never come into its own precisely because it is deprived not just of the means of stable living but also of the means of producing its own representation. Lacking a shared language and social compass, riven by fissiparity, its members do not flock to support far-rightist parties so much as disperse and drop out of the voting game altogether as from other forms of civic participation.

A Bourdieusian framework for the comparative sociology of urban inequality

Urban Outcasts sketches a historical model of the ascending regime of poverty in the city at the turn of the century. It forges notions – ghetto, hyperghetto, antighetto, territorial stigmatisation, advanced marginality, precariat – geared to developing a comparative sociology of relegation capable of eschewing the uncontrolled projection across borders of the singular experience of a single national society tacitly elevated to the rank of analytic benchmark. It does so by applying to urban questions five principles undergirding Pierre Bourdieu's approach to the construction of the sociological object. These principles are worth spotlighting by way of closing since this is a facet of the book that has been overlooked even by its more sympathetic critics.[19]

The first principle derives directly from 'historical epistemology', the philosophy of science developed by Gaston Bachelard and Georges Canguilhem, and adapted by Bourdieu for social inquiry: clearly demarcate folk from analytic notions, retrace the travails of existing concepts in order to cast your own and engage the latter in the endless task of rational rectification through empirical confrontation (Bourdieu, Chamboredon, & Passeron, [1968] 1991; Broady, 1991). Such is the impulse behind the elaboration of an institutionalist conception of the ghetto as a Janus-like contraption for ethno-racial enclosure, commenced in *Urban outcasts* and completed in its sequel, *The two faces of the ghetto*, which further differentiates the ghetto from the ethnic cluster and the derelict district; compares it with its functional analogues of the reservation, the camp and the prison; and stresses the paradoxical profits of ghettoisation as a modality of structural integration for the subordinate population (Wacquant, 2015). Second, comes the relational or topological mode of reasoning, deployed here to disentangle the mutual connections and conversions between symbolic space (the grid of mental categories that orient agents in their cognitive and conative construction of the world), social space (the distribution of socially effective resources or capitals), and physical space (the built environment resulting from rival efforts to appropriate material and ideal goods in and through space).

The third principle expresses Bourdieu's radically historicist and agonistic vision of action, structure and knowledge: captures urban forms as the products, terrains, and stakes of struggles waged over multiple temporalities, ranging from the *longue durée* of secular constellations to the mid-level tempos of policy cycles to the short-term phenomenological horizon of persons at ground level. In this perspective, America's Black Belt and France's Red Belt, like districts of relegation in other societies, emerge as historical animals with a birth, maturity, and death determined by the balance of forces vying over the meshing of class, honour, and space in the city. Similarly, the hyperghetto of the US metropolis and the anti-ghettos of Western Europe are not eternal entities springing from some systemic logic but time-stamped configurations whose conditions of genesis, development, and eventual decay are sustained or undermined by distinct configurations of state and citizenship. The fourth tenet recommends the use of ethnography as an instrument of rupture and theoretical construction, rather than a simple means for producing an experience-near picture of ordinary cultural categories and social relations. It implies a fusion of theory and method in empirical research that overturns the conventional division of intellectual labour in urban inquiry marked by the routine divorce of microscopic observation and macroscopic conceptualisation.[20]

Last but not least, we must heed the constitutive power of symbolic structures and track their double effects, on the objective webs of positions that make up institutions, on the one side, and on the incarnate systems of dispositions that compose the habitus of agents, on the other. As illustrated by territorial stigmatisation, this principle is especially apposite for the analysis of the fate of deprived and disparaged populations, such as today's urban precariat, that have no control over their representation and whose very being is therefore moulded by the categorisation – in the literal sense of *public accusation* – of outsiders, chief among them professionals in authoritative discourse such as politicians, journalists, and social scientists. This is to say that the sociologist of marginality must punctiliously abide by the imperative of epistemic reflexivity and exert constant vigilance over the myriad operations whereby she produces her object, lest she gets drawn into the classification struggles over districts of urban perdition that she purports to objectivise.

These five principles propel the comparative dissection of the triadic nexus of class (trans)formation, graduations of honour, and state policy in the nether regions of metropolitan space across the Atlantic presented in this book. They can also fruitfully guide a triple extension of the sociology of urban relegation in the era of social insecurity across continents, theoretical borders, and institutions. Geographically, they can steer the adaptation of the schema of advanced marginality via sociohistorical transposition and conceptual amendment to encompass other countries of the capitalist core as well as rising nations of the Second world where disparities in the metropolis are both booming and shape-shifting rapidly.[21] Theoretically, taking Bourdieu's distinctive concepts and propositions into city trenches offers a formidable springboard to both challenge and energise urban sociology *in globo* (Wacquant, 2018). It does not just add a new set of powerful and flexible notions (habitus, field, capital, doxa, symbolic power) to the panoply of established perspectives: it points to the possibility of reconceptualising the urban as the domain of accumulation, differentiation, and contestation of manifold forms of capital, which effectively makes the city a central ground and prize of historical struggles.

On the institutional front, the consolidation of a new regime of urban marginality begs for a focused analysis of the policy moves whereby governments purport to curb, contain or reduce the very poverty that they have paradoxically spawned through economic 'deregulation' (as re-regulation in favour of firms), welfare retraction and revamping, and urban retrenchment. It calls, in other words, for *linking changing forms of urban marginality with emerging modalities of state-crafting*. This is done in my book *Punishing the Poor*, which enrols Bourdieu's concept of bureaucratic field to diagram the invention of a punitive mode of regulation of poverty knitting restrictive 'workfare' and expansive 'prisonfare' into a single organisational and cultural mesh flung over the problem territories and categories of the dualising metropolis (Wacquant, 2009b, 2012).[22] The wards of urban dereliction wherein the precarised and stigmatised fractions of the post-industrial working-class concentrate turn out to be the prime targets and testing ground upon which the neo-liberal Leviathan is being manufactured and run in. Their study is therefore of pressing interest, not just to scholars of the metropolis, but also to theorists of state power and to citizens mobilised to advance social justice in the twenty-first-century city.

Notes

1 First published as Wacquant L. (2016). Revisiting territories of relegation: Class, ethnicity and state in the making of advanced marginality. *Urban Studies*, 53(6), 1077–1088. Funding: The MacArthur Foundation, the W. E. B. Du Bois Institute at Harvard University, and a Leverhulme Trust grant for the international network on 'Edgework: Comparative Studies in Advanced Urban Marginality' provided support for this research.
2 For an account of the biographical, analytic, and civic underpinnings of this project, see Wacquant, 2009a, esp. pp. 106–110.
3 A historical recapitulation of the loaded meanings and persistent ambiguities of the notion of 'community' in US history is Bender (1978).
4 The mutual contamination and common intermingling of scholarly and ordinary visions of urban life is stressed by Hall, 1988 and Low, 1996.
5 This parallel 'black city within the white' is depicted by St. Clair Drake and Horace Cayton in their classic study, *Black Metropolis: A Study of Negro life in a Northern City* (Drake & Cayton, [1945] 1993).
6 This spatial and social differentiation, leading to contest and confrontation over the norms and fate of the 'neighbourhood', is skilfully documented in the work of the preeminent sociologist of black America of her generation, Patillo (2000, 2007).
7 This is evidenced by the confused announcement by Prime Minister Lökke Rasmussen of a 2010 government plan to 'confront the parallel societies of Denmark' by targeting 29 officially designated 'ghettos', defined by the confounding combination of immigration, joblessness, and crime (see Rasmussen, 2010, pp. 1–7, 37–39).
8 The protean cultural production of the city underbelly or underworld as the 'accursed share' of urban society is dissected by Dominique Kalifa (2013).
9 On the Danish case, see Skifter (2010).

10 Collapsing these three levels conflates collective conscience with social morphology, elite discourse and everyday action, and mechanically leads to overestimating both the novelty and the potency of ethnicity as a determinant of life-chances, as does Amselle (2011).
11 A stimulative reinterpretation of this characterisation is Banton (2007).
12 For a model study breaking down ethnicity across social forms and scales (see Brubaker, Feischmidt, Fox et al., 2008); a germane argument from an analytic angle is Wimmer (2013).
13 Curiously, this thesis has gone virtually unnoticed in the extended symposia devoted to *Urban outcasts* by the journals *City* (December 2007 and April 2008), *International Journal of Urban and Regional Research* (September 2009), *Revue française de sociologie* (December 2009), *Pensar* (Winter 2009), and *Urban Geography* (February 2010), which have moreover concentrated either on the diagnosis of the black ghetto or on the evolution of the French/European periphery at the cost of scotomising the book's comparative agenda.
14 For a varied panorama, Shipler (2004); Clerc (2004); Andress & Lohmann (2008). The Danish case is examined by Hansen (2010). Revealingly, the US-inspired category of the 'working poor' was introduced into French official statistics in 1996, in European Union statistics in 2003, and in German government reports in 2009.
15 The built-in blindness of such research to macrostructural economic and political forces is stressed by Slater (2013).
16 See the articles and the wide-ranging bibliography gathered by Slater, Pereira and Wacquant (2014). An extension to Denmark is Qvotrup Jensen and Christensen (2012); see also Sernhede (2009).
17 In April 2011, the High Council for Fighting Discrimination and for Equality (HALDE) recommended to the French government that residential location be added to the 18 criteria on the basis of which national labour law sanctions discrimination, in recognition of the prevalence of 'address discrimination'.
18 For a demonstration covering the 29 areas officially designated as '*ghetto- område*' – which conveniently obscures the fact that they are simply '*forsømt*' (dilapidated) – in Denmark (see Schultz Larsen, 2011, pp. 47–67).
19 For a signal exception, see Kristian Delica (2011). These principles are explicated and exemplified in Bourdieu and Wacquant (1992).
20 The peculiar genre of research unthinkingly labelled 'urban ethnography' in the English-speaking academy is blissfully atheoretical, as if one could carry out embedded observation of anything without an orienting analytic model, while grand theories of urban transformation show little concern for how structural forces imprint (or not) patterns of action and meaning in everyday life. One of the aims of *Urban outcasts* is to bridge that chasm and to draw out the manifold empirical and conceptual benefits arising from continual communication between field observation, institutional comparison and macroscopic theory.
21 An amplification across the Channel is offered by Atkinson, Roberts, and Savage (2012), partial adaptations to South Africa, Brazil, and China, respectively, are Murray (2011), Perlman (2010), and Wu and Webster (2010). See also the diverse works of the scholars affiliated with the interdisciplinary network at www.advance durbanmarginality.com.
22 For an analysis of the international diffusion of the penalisation of poverty as a core component of neo-liberal policy transfer, see Wacquant (2009c).

References

Amselle, J.-L. (2011). *L'ethnicisation de la France*. Fécamp: Nouvelles Éditions en Lignes.

Andress, H.-J., & Lohmann, H. (Eds.) (2008). *The working poor in Europe: Employment, poverty, and globalization*. Cheltenham: Elgar Publishing.
Atkinson, W., Roberts, S., & Savage, M. (Eds.) (2012). *Class inequality in austerity Britain: Power, difference and suffering*. Basingstoke: Palgrave Macmillan.
Banton, M. (2007). Max Weber on 'ethnic communities': A critique. *Nations and Nationalism*, 13(1), 19–35.
Bender, T. (1978). *Community and social change in America*. New Brunswick: Rutgers University Press.
Bourdieu, P. (1990). *Language and symbolic power*. Cambridge: Polity Press.
Bourdieu, P., & Wacquant L. (1992). *An invitation to reflexive sociology*. Chicago, IL: University of Chicago Press.
Bourdieu P., Chamboredon J.-C., & Passeron J.-C. [1968] (1991). *The craft of sociology: epistemological preliminaries*. New York; Berlin: Walter de Gruyter.
Broady, D. (1991). *Sociologi och epistemologi. Om Pierre Bourdieus sociologi och den historiska epistemologin*. Stockholm: HLS Förlag.
Brubaker, R., Feischmidt, M., Fox, J., & Grancea, L. (2008). *Nationalist politics and everyday ethnicity in a Transylvanian town*. Princeton: Princeton University Press.
Cingolani, P. (2011). *La précarité*. 3rd ed. Paris: PUF.
Clerc, D. (2004). *La France des travailleurs pauvres*. Paris: Grasset.
Delica, K. (2011). Sociologisk refleksivitet og feltanalytisk anvendelse af etnografi: om Loïc Wacquants blik på urban marginalisering. *Dansk Sociologi*, 22(1), 47–67.
Drake, St. C., & Cayton, H. [1945] (1993). *Black metropolis: A study of Negro life in a northern city*. Chicago, IL: University of Chicago Press.
Goffman, E. (1964). *Stigma: Notes on the management of spoiled identity*. Englewood Cliffs: Prentice-Hall.
Hall, P. (1988). *Cities of tomorrow: An intellectual history of urban planning and design in the twentieth century*. Oxford: Basil Blackwell.
Hansen, F. K. (2010). *Fattigdom i EU-landene – og dansk fattigdom i europærisk perspektiv*. Copenhagen: CASA.
Kalifa, D. (2013). *Les bas-fonds. Histoire d'un imaginaire*. Paris: Seuil.
Kalleberg, A. L. (2011). *Good jobs, bad jobs: The rise of polarized and precarious employment systems in the United States, 1970s–2000s*. New York, NY: Russell Sage Foundation.
Liogier, R. (2012). *Le mythe de l'islamisation. Essai sur une obsession collective*. Paris: Seuil.
Low, S. (1996). The anthropology of cities: Imagining and theorizing the city. *Annual Review of Anthropology*, 25, 383–409.
Massey, D., & Denton, N. (1993). *American apartheid: Segregation and the making of the underclass*. Cambridge, MA: Harvard University Press.
Murray, M. J. (2011). *City of extremes: The spatial politics of Johannesburg*. Durham: Duke University Press Books.
Maurin, L., & Savidan, P. (2008). *L'État des inégalités en France 2009. Données et analyses*. Paris: Belin.
Perlman, J. (2010). *Favela: Four decades of living on the edge in Rio de Janeiro*. New York, NY: Oxford University Press.
Pan Ké Shon, J.-L., & Wacquant, L. (2012). *Le grand hiatus: Tableau raisonné de la ségrégation ethnique en Europe*. Paper presented at the Journée INED on 'La ségrégation socio-ethnique: Dynamiques et conséquences', Institut national d'études démographiques, Paris, France, 13 June. Retrieved from www.ined.fr/en/news/

scientific-meetings/seminaries-colloque-ined/la-segregation-socio-ethnique-dynamiques-et-consequences/.
Pattillo, M. (2000). *Black picket fences: Privilege and peril among the black middle class*. Chicago, IL: University of Chicago Press.
Pattillo, M. (2007). *Black on the block: The politics of race and class in the city*. Chicago, IL: University of Chicago Press.
Pelizzari, A. (2009). *Dynamiken der Prekarisierung. Atypische Erwerbsverhältnisse und milieuspezifische Unsicherheitsbewältigung*. Konstanz: UVK Verlag.
Qvotrup Jensen, S., & Christensen, A.-D. (2012). Territorial stigmatization and local belonging. *City, 16*(1–2), 74–92.
Rasmussen, L. (2010). *Ghettoen tilbage til samfundet – et opgør med parallelsamfund i Danmark*. Copenhagen: Regeringen.
Schultz Larsen, T. (2011). Med Wacquant i det ghettopolitiske felt. *Dansk Sociologi, 22*(1), 47–67.
Sernhede, O. (2009). Territorial stigmatisering: Unges uformelle læring og skolen i det postin- dustrielle samfund. *Social Kritik, 118*, 5–23.
Shipler, D. K. (2004). *The working poor: Invisible in America*. New York, NY: Knopf.
Skifter, A. H. (2010). Spatial assimilation in Denmark: Why do immigrants move to and from multi-ethnic neighbourhoods? *Housing Studies, 25*(3), 281–300.
Slater, T. (2013). Your life chances affect where you live: A critique of the 'cottage industry' of neighbourhood effects research. *International Journal of Urban and Regional Research, 37*(2): 367–387.
Slater, T., Pereira, V., & Wacquant, L. (Eds.) (2014). Special issue on 'territorial stigmatization in action'. *Environment & Planning D, 46*(6): 1263–1402.
Standing, G. (2011). *The precariat: The new dangerous class*. London: Bloomsbury.
Wacquant, L. (2008a). *Urban outcasts: A comparative sociology of advanced marginality*. Cambridge: Polity Press.
Wacquant, L. (2008b). Ghettos and anti-ghettos: An anatomy of the new urban poverty. *Thesis Eleven, 94*(1), 113–118.
Wacquant, L. (2009a). The body, the ghetto and the penal state. *Qualitative Sociology, 32*(1), 101–129.
Wacquant, L. (2009b). *Punishing the poor: The neoliberal government of social insecurity*. Durham and London: Duke University Press.
Wacquant, L. (2009c). *Prisons of poverty*. Minneapolis, MN: University of Minnesota Press.
Wacquant, L. (2011). A Janus-faced institution of ethno-racial closure: A sociological specification of the ghetto. In R. Hutchison R. & B. Haynes (Eds.), *The Ghetto: contemporary global issues and controversies* (pp. 1–31). Boulder, CO: Westview.
Wacquant, L. (2012). Crafting the neoliberal state: Workfare, prisonfare and social insecurity. *Sociological Forum, 25*(2), 197–220.
Wacquant, L. (2015). *The two faces of the Ghetto*. New York, NY: Oxford University Press.
Wacquant, L. (2018). Bourdieu comes to town: pertinence, principles, applications. *International Journal of Urban and Regional Research, 42*(1), 90–105.
Wilson, W. J. (1996). *When works disappears: The world of the new urban poor*. New York, NY: Knopf.
Wimmer, A. (2013). *Ethnic boundary making: Institutions, power, networks*. New York, NY: Oxford University Press.
Wu, F., & Webster, C. (Eds.) (2010). *Marginalization in urban China: Comparative perspectives*. New York, NY: Palgrave Macmillan.

2 State-making as space-making

The three modes of the production of space in Istanbul

Sinan Tankut Gülhan

Introduction

This chapter proposes a new approach to make sense of the social and spatial transformations in Istanbul by suggesting a tri-modal perspective for understanding urban change. The recent history of modernisation in Istanbul was primarily shaped by the interplay of the state and space, while mediated by the equally powerful structures of the world economy, symbolic power, politics, and habitus. Here, I delineate three different modes of state-spaces – namely, *world-imperial-quasi-colonial*, *etatist/national developmentalist*, and *neo-liberal-state-corporate-alliance* modes – that determined the historical trajectory of Istanbul in the last century and a half and the intermingling of and successive antagonisms incumbent upon these three modes that shaped contemporary Istanbul.

First, I will define the main components of a state-space analysis and show how a historico-geographical imagination asks for a return to an often-overlooked structural phenomenon known as the state. Second, I will define my theoretical approach by explaining the four layers – objectifying instruments – of analysis. Based on the conceptual framework developed by Bob Jessop, Neil Brenner, and Martin Jones (2008) I will explain that territory, place, scale, and networks can be employed with slight modifications to analyse urban transformation in Istanbul. In the third part, I locate the three modes of state-space in Istanbul and follow their linkages, continuities, and ruptures. As the continuities are emphasised, it is important to underline the fact that these three modes are not mutually exclusive and different formulations and representations of each mode had come to play a significant role in successive re-arrangements.

Bringing the state back to the urban space

Grappling with space requires a study engrained in both concrete and abstract formulations of the state as well as an analysis that is reflexive of the spatio-temporal spectrum. For urbanists up to the 1960s, the urban question was first and foremost a question of the natural, or cultural, habitat. For the Chicago School, the city was undoubtedly a laboratory to gauge human-animal spirits and in this laboratory; the state's agency is merely defined by the element of social control (Park & Burgess, 1984). Amos Hawley, the renowned mid-twentieth-century sociologist who salvaged human ecology from the ruins of evolutionary naturalism, redefined the task of urban ecology as research on the static and dynamic aspects of collective morphology (Hawley, 1944, p. 404). The founding figures of urban scholarship, the Chicago School and their post-war quantitative brethren came to be recognised with their steadfast denial of the supply side dynamics in the processes of making space (Gottdiener, 1985, pp. 25–69; Saunders, 1981, pp. 52–84).

The new urban sociology of the late 1960s brought back the state into the picture and entered the scene with a bang. Consciously described as a moment of rupture in conceptualising the urban space, Manuel Castells heralded the foundation of an urbanistic science based on Marxism. When Castells sought an answer to the crucial question of if there is an actually existing urban sociology, he provided his own answer: no, as long as urban sociology lacked a scientific object, which, he claimed, there was none. Castells lamented that previous schools of urban sociology were lacking in their endeavour to define an object. And unless an object is defined in terms of a scientific research programme there would be nothing to speak of scientific endeavour (Castells, 1976a, 1976b). He argued that the city can only be distinguished by its special place in accumulation and circuits of capital; it is the site of collective consumption. Collective consumption did not just mean municipal services like the garbage collection and disposal, sewage, transportation, and water infrastructure, but it covered the whole array of capitalist reproduction of labour power – education, coercion, such as social control, housing, family, urban renewal, and so forth and was the wellspring of social movements (Castells, 1977). It is no wonder that the first issue of the *International Journal of Urban and Regional Research*, the flag bearer of new urban sociology, was on the Marxist conception of the state. Castells' work in the latter part of the 1970s was coupled with the prolific and imaginative research conducted by Ray E. Pahl, Jean Lojkine, and Edmond Préteceille in a wider research programme that entailed a thorough examination of the state effects in the city (Castells, 1978; Lojkine, 1976, 1977; Pahl, 1977; Préteceille, 1981).

However, with the onset of neo-liberalism instigated by the Thatcherist-Reaganist political and economic reorientation of the early 1980s (Harvey, 2005) and its reverberations throughout the world in different roles and concoctions, which Brenner, Peck, and Theodore call variegated neo-liberalism (Brenner, Peck, & Theodore, 2010), the interest in the role played by the state in the production of space withered. In its stead, newspeak of globalisation captured the imagination of urbanists and social scientists alike from the early 1990s onward (Bourdieu & Wacquant, 2001). Only recently has there been a reinvigorated interest in how political power and urbanisation are related.

The coming of the sole logical consequence of neo-liberalism, that is, populism, instigated a reformulation of ready-made explanations where the state was not merely withering. Such questioning brought forward a rehashing of well-worn discussions such as the nature of the state, the social, cultural, and ideological transmutations of the state, and its processual formulation under a higher level of abstraction – *statehood*. Furthermore, a relatively new mode of inquiry grounded in spatial analysis and critical urbanism invoked and laid bare the need for a critical examination towards reinvigorating space as an element of sociological study, providing an ability to better grasp the socio-spatial dialectic. A novel conceptualisation of the relationality between the state and space, what Brenner (1999a, 2004) defined as *state spatiality*, provides an elaborate social, geographical, and historical imagination – in addition to conceptual tools of explanation – while raising deeper, cross-cutting, and at times, paradoxical questions.

Undoubtedly, there is a need to bring back the state into research on space. Yet, the path to an adequate interpretation of the state as an active agent in the formation of the built environment is laden with obstacles. This necessarily implies a *state spatiality* that is continuously reproduced via the flux of global capital accumulation. Hitherto the state and space, the two proposed axes of a processual investigation, have been studied either as objective parallels with clear-cut relationships of causality, or as two non-intersecting reified entities. Neil Brenner has brilliantly summarised the arguments regarding the state-space nexus and stated that three assumptions are the underlying factors in the conceptual separation of the state and space: spatial fetishism, methodological territorialism, and methodological nationalism. These three in effect shaped the hitherto prevalent state-centric epistemology in social sciences (Brenner, 2004, pp. 37–47). Brenner argued that 'state-centric epistemologies freeze the image of national state territoriality into a generalized feature of social life' (Brenner, 2004, p. 43).

State, therefore – as per Brenner's criticism – was deemed to be an externality to space. In the eyes of the twentieth century Western Marxists the state was ontologically instrumental, an extension of class domination that represented nothing but a baton that regulates social relationships of production in order to continuously produce consent (Soja & Hadjimichalis, 1979; Soja, 1980, 1989). For others, the state was an apparatus of redistribution, a wedge to extend the functional operations of different social strata, a neutral arbiter, and the end result of the pluralistic political processes. Yet, such perceptions of the state and space had for a long time heavily underestimated the multitudinous depths of their relationality, overlooked the possibilities of the state to materially, and mentally, mould new spatial forms that reflect the state's immanent conflicting qualities, its incipient, and at times, fettered developmental tendencies, and its vicissitudinous role in the capital circuits.

Formulating the 'social space' in the territory, place, scale, and network (TPSN) framework

In naming these modes, I tried to convey the sense that the first part refers to the politico-ideological views prevalent in the historico-geographical periodisation, while the second part represents the dominant economic mode of accumulation.

The amalgamation of historico-geographical periodisation and economic mode of accumulation bring together four different layers of spatial analysis. Bob Jessop, Neil Brenner, and Martin Jones pointed out that under the circumstances of capitalist relations of production, the key to theorising the space and state nexus involves four levels: territory, place, scale, and networks (Jessop, Brenner, & Jones, 2008). The TPSN framework is invaluable in its contribution to the sociological study of urban matters; equally important was the criticism of the conceptual rigidities and the invitation to a strategic-relational approach. Yet, I would like to re-imagine the quadruple in the humblest of hopes that it served better in understanding the development of urban phenomena in Istanbul and suggest to reformulate three of the layers in order to better describe the spatial transformation from a historico-geographical perspective: practical/lived spaces as territory, places as representations of space, and finally networks as patterns of symbolic exchanges that defined the structures of meanings embedded in particular doxa, that is habitus.

Territory is perhaps the easiest to conceptualise, as it envelopes 'bordering, bounding, parcelisation, enclosure' while it produces the 'construction of inside/outside divides' (Jessop, Brenner, & Jones, 2008, p. 393). It is the most physical layer of socio-spatial structuration; hence, I would argue that it closely echoes Henri Lefebvre's first moment in spatial analysis: the practical space that is perceived, that 'embodies a close association (...) between daily reality (daily routine) and urban reality (the routes and networks which link up the places set aside for work, "private" life and leisure)' (Lefebvre, 1991, p. 38). Hence, the territory is not only about juridico-political boundaries, such as state lines or city limits, or geographical signifiers like highways, railways, industrial wastelands, brownfields, parks, shopping malls, and so forth, but contains practical relations that are produced within the realm of everyday life.

Place, on the other hand, as Jessop, Brenner, and Jones point out is frequently subject to a fallacy 'which treats places as discrete, more or less self-contained, more or less self-identical ensembles' (Jessop, Brenner, & Jones, 2008, p. 391). Perhaps, place, as a sociological concept, is immune to objectification and in employing place as an element of socio-spatial analysis a plethora of nostalgia-laden faux historicism, subjective fallacies, and characterisations based on pseudo-realities or illusions are unavoidable. I would like to suggest that places are conscious reconstructions of real relationships between individuals and groups, and thus 'place' as a mental space; a cognitive map that relays modes of conduct, language, and gender roles. Here, the stick can be bent both ways, Pierre Bourdieu's work on social space from his earliest days as a field ethnographer in Algeria to his inheritor Loïc Wacquant's fruitful intervention in urban theory, primarily theorised place as a mental space (Bourdieu, 1970, 1977, 1984; Wacquant, 2018). On the other end of the spectrum, Henri Lefebvre's trialectics of space dwelt on the crucial concept of representations of space, wherein 'ideology and knowledge' are combined 'within a (social-spatial) practice' and 'their intervention occurs by way of construction (...) by way of architecture' (Lefebvre, 1991, pp. 42, 45). Bourdieu's and Lefebvre's notions of space can be rightfully posed as counter-arguments, although in my view their continuities and similarities are much more prominent and meta-theoretical doxasophy weighs down the possible interplay of their spatial imagination. In a nutshell, the place is the mental construct par excellence; it is the ultimate embodiment of the representations of space.

Scale, the third category, is perhaps the most contentious of all (Brenner, 2001; Marston, 2000; Marston & Smith, 2001). However, it is the most well thought out component of the TPSN framework. Brenner wrote a series of essays on reworking the notion of scale as an element of state-space analysis (Brenner, 1999a, 1999b, 1997) and developed a novel understanding of 'scalar structuration' (2001) which enabled a conceptualisation of the multi-scalar investigation of state-spaces beyond national borders (2004, p. 74). Sallie Marston called for a theorisation of scale construction towards social reproduction or consumption (Marston, 2000, pp. 238–239). Both arguments have their own merits. Yet, my analysis here grapples with the 'multiple spatialities of scale' and deals with how different scales 'constitute geographies and choreographies of social power' (Brenner, 2001, pp. 606–608).

Networks, the fourth category, require an expanded discussion on the state of the art in contemporary social theory. Yet, I would rather shortly summarise a certain vein of criticism here. We need to disabuse ourselves from the notion that the networks are a novel conceptual tool that helps us explain social phenomena. The ubiquity of the term belies its usefulness and unless one is really into computer sciences, networks have always been a part of human social interactions. Networks do not construct fields; they are apt metaphors, but, sociologically speaking, their usefulness as an objectifying conceptual set is limited. Instead, we need to dig through the meanings ensconced within networks. In other words, how 'the production of a commonsense [sic] world endowed with the *objectivity* secured by consensus on the meaning (*sens*) of practices and the world' (Bourdieu, 1977, p. 80) can be pertinently located carries a much urgent implication.

These four levels, territory/practical space, representations of space, scale, and habitus, made themselves felt in three different modalities of state-spaces in Istanbul. My intention in putting a slightly revised TPSN schema to work, as per Jessop, Brenner, and Jones, is 'to decipher the strategies and tactics of individual and collective agents, organisations, and institutions that are engaged in contentious politics' as well as 'to pose new questions regarding the interplay between the spaces of contentious politics and the geo-historical periodisation of capital accumulation and state power' (Jessop, Brenner, & Jones, 2008, p. 398). *Habitus* is perhaps the only novel contribution to the TPSN framework and since every social space has its own *habitus*, its own symbolic order, and its own hierarchy of meanings it is necessary to locate it in the dynamics of state-spaces.

Mode I: world-imperial-quasi-colonial state-spaces, 1839–1920s

In the first mode, the dissolution of a world empire and the pangs of incorporation to a world economy emerged with its particular state-spaces. A caveat is in order, world empire does not carry the same connotation as the term imperial or imperialism implies; and as Immanuel Wallerstein explained in great detail, the politico-institutional entity of the world empires preceded the formation of the world economy as well as the modern world system (Hopkins et al., 1987; Wallerstein, 1974, pp. 60–63; Wallerstein, Decdeli, & Kasaba, 1987). A world empire is a primitive means of economic domination (Wallerstein, 1974, p. 15) and a stark contrast exists between a world economy, which is primarily market-oriented and profit-driven, and a world empire, which is decisively political. Wallerstein emphasised that up until the emergence of the modern world economy in the seventeenth century world-economies almost always transformed into world empires (1974, pp. 16, 311). Hence, late Ottoman Istanbul was forged politically in the symbolic and ideological field of a world empire.

Territorial breaking up of the empire surfaced in Istanbul as the old, intra-muros Istanbul gradually faded away and gave way to the northern expansion of the city towards Pera and today's Beşiktaş. The representations of space were opulently rich in this mode since place-making of the time involved rebuilding and replicating European capitals in smaller scales, with an arabesque accent, as the new imperial palaces showed. Photography provided a novel, and technological, shortcut to strategically reproduce the urban structures. Abdulhamid II, although very seldom photographed himself, used photography extensively to gather information on his vast empire (Allen, 1984; Shaw, 2009).

The scale of this state-space modality had to deal with the powerful forces of centrifugal provincialism (Mardin, 1973) and centripetal incorporation to the world economy (Hopkins et al., 1987; Keyder, 1987; Wallerstein, Decdeli, & Kasaba, 1987). Istanbul as the capital of a world empire had its own peculiar scalar characteristics, wherein a vast hinterland, and a nodal concentration of surplus produce made the city unrivalled within the realm of the world empire up until its incorporation to the modern world system. The political economy of that multi-scalar mismatch sowed the seeds for what Göran Therborn explained as 'reactive modernisation' (Therborn, 2017, pp. 7–32, 147–164). From an economic perspective, by the end of the nineteenth century, the empire was colonial and its commercial hub was the Galata (Pera) district of Istanbul. Yet, the juridico-political structure was relatively autonomous and skilfully harnessed the anti-Western emotions prevalent in the Muslim provinces.

From this chasm was born a habitus of a new bureaucratic elite class. The unfettered Westernisation was a source of ridicule from the beginning among this bureaucratic elite class (Mardin, 1995, pp. 21–65). The habitus of the world empire – the Ottoman *millet* system – were rapidly replaced by new attempts in establishing an enlightened military-bureaucratic class obedient to the Ottoman central bureaucracy. Even the critics of the imperial absolutist monarchy sought not for more Westernisation but appealed to the Islamic identity and later to the Turkishness of the nation. The scarred habitus was a product of the devastation brought on by the loss of the Balkans and it was inimical to the Western values. *Mahalle* – the Turkicised Arabic word for a district – came to represent the authentic Muslim community where traditional values were respected (İnalcık,

Lived space/territory	Representations of space/place-making
State spaces of the Bosphorus' Western shores Pera as a Western commercial colony	*Galata vs. Süleymaniye* Extending old bourgeois neighbourhoods Photography

Scale	Habitus
Economic colonialism Juridico-political absolutism Provincial backlash	*Mahalle-qua-the-State* Big landowners and merchant capitalists Ethnic and *faith-based* distribution of power and resources

Figure 2.1 State-space mode I: world-imperial-quasi-colonial.

1990, 2015, p. 143). This ethnic and faith-based spatial segregation would play a decisive role in the successive state-space modalities as the birthplace of both Islamist and socialist political movements in the twentieth century.

Mode II: etatist/national developmentalist state-spaces

The second mode of state-space in Istanbul owes its emergence to the closely guarded privileges of the military-bureaucratic class; this time, however, under a novel disguise: Kemalism. The features of Kemalist modernism only began to be studied in the 1960s and 1970s. In this period, Turkey witnessed new and internationally comparative interest in fertile discussions on the foundational characteristics of the Turkish state. The relationship between the state and society, contradictions between different classes and the roles they played in the state strategies, had found receptive audiences within academic circles. The delayed modernisation, the maladroit national developmentalist strategies that are subject to the vagaries of the ebb and flow of Cold War politics, the stunted development of underdevelopment, the late incorporation to the world economy, the tension between centripetal and centrifugal forces that played out in the political and ideological antagonism of centre and periphery, and the unravelling of this tension at the realm of politics, played a prominent role in efforts to grasp the uniqueness of the state and society relations in Turkey (Boratav, 1981; Boratav, Keyder, & Pamuk, 1984; Keyder, 1976, 1983; Mardin, 1971, 1973, 1995).

The territorial aspect, the use of practical space depended on the dispossession of non-Turkified minorities, the wiping away of their traces on the newly established country's territory, and most important of them all, in re-territorialising Istanbul's urban land. A newly found centralised power of the state helped usher in a period of toponymy in Istanbul: Pera became Beyoğlu, Tatavla turned into Kurtuluş and so forth.

The peasants were wilful partners in this new territorial expansion of the state-space, as long as they subscribed to the predominant ideological apparatuses of the state, their massive appropriation of space was overlooked, or implicitly encouraged. From that conscious state manoeuvre, the *gecekondus* – literally meaning dropped in a night – the Turkish slums arose. Very similar to what Asef Bayat described as the 'quiet encroachment of the ordinary' (Bayat, 1997, pp. 4–17), beginning from the first freely contested election in 1946, Istanbul and its hitherto pristine public land became a terrain of silent politics.

However, the quiet encroachment was pragmatically organised. Time and again, central and municipal governments declared their intention to wipe out the slums. On the other hand, places of the republican identity permeated, bolstered, and strengthened the military-bureaucratic elite's claim on power and their imagined historiography. The scalar needs of this second state-space modality were subordinated by a national developmentalist industrialisation drive. The more the merrier was the motto, and the city went beyond its historical borders, remaking itself in the image of an insuperable growth machine. The machinations of scales have been well scrutinised: clientelism, ethnically and religiously motivated hometown connections did not only survive intact, but they found a new lease on political, cultural, and social influence by dint of local level redistribution of urban resources in the hands of the state (Aslan, 2004; Ayata, 2008; Erman, 2001).

The representations of space were nationalistic in appearance. From the 1930s on, the Kemalists turned their attention to Istanbul. Even though the leading architectural and planning historians recognised a lopsided interest in Ankara in the early republican period (Tekeli, 1994; Kuban, 2000; Bozdoğan, 2001; Sargın, 2002; Gül, 2009), the only professional architectural journal of the era, *Arkitekt/Mimar*, and the pages of the best-selling newspaper *Akşam* were awash with speculations on the fate of modernist planning in Istanbul. For the first time in its life since the fourth century, Istanbul found itself secondary to another capital, Ankara. Yet, Henri Prost's *beaux-arts* inspired modern planning efforts did not come to fruition due to World War II (Bilsel, 2011; Gül & Lamb, 2004; Tekeli, 1994).

Traditionally the Turkish state had functioned in a national developmentalist mentality with a curious twist. Quite unlike the Fordist-Keynesian response to the housing problem in the West, that either promulgated the state as the chief financial regulator of the housing market by means of mortgages or through actively pursuing building housing for the socially disadvantaged (Ball, 1981; Checkoway, 1980; Dagnaud, 1983; Goursolas & Atlas, 1980), the Turkish state's response to the housing question was much more passive. Especially as a result of the 1960s policy-making decisions, a non-interventionist planning programme was initiated by the central planning authorities of the Turkish state. As a consequence, in the 1960–1980 period, 70 per cent of housing investment was in private hands (Batuman, 2006; Keleş, 2010). The state did not hesitate in undertaking grandiose schemes of infrastructural projects, irrigation canals, hydroelectric power plants, and highways while marginally involved with the housing market. The minuscule state contribution in urban development was mainly oriented towards the provision of heavily subsidised residence construction for the swelling ranks of apparatchiki such as teachers, officers, university professors, and doctors, cross-cutting the social landscape under the mantle of unquestioned allegiance to state bureaucracy (Keyder, 1987, pp. 47–48).

The habitus in Istanbul was a product of this choice and fed into the polarisation of Istanbul's urban landscape. The state professionals, Kemalist elites, enjoyed limited privileges, while the *gecekondu* dwellers, the migrants, barely found possibilities of survival. The apartment buildings replaced the low-density housing in the inner city. Thus, apartment complexes served as the beacons of an urban way of life in Istanbul. The loyalty of the apparatchik or old bourgeois residents was not only solidified through housing subsidies and collective consumption but also their self-reflexive consciousness of vanguardist Kemalism had contributed dearly to what some described as the core-periphery conflict in Turkish politics. A borrowed word from French, the 'lojman' – *logement* – had long created interesting residential segregation, an archetypal gated community in the Turkish urban landscape.

Yet, on the other side of the territorial divide, agony and deprivation persisted. The last *cholera* epidemic in Istanbul broke out in 1970 in the slum neighbourhood of Sağmalcılar and claimed more than 100 lives (Gürsoy, 1970; İpekçi, 1970). Out of this misery, new cultural forms were born in the slums of Istanbul's periphery. *Popular music* was defined by *arabesk*, a fusion of Egyptian and folkloric music with fatalistic lyrics (Özbek, 2013; Stokes, 1998), *Turkish movie industry*, as one of the two survivors of worldwide

Lived space/territory	Representations of space/place-making
Industrialisation	Novelists' socialist realism
Dispossession (and re-possession) of non-Muslim owned inner-city land	'Dangerous classes'
	The periphery
Quiet encroachment on public lands	Extending and re-inventing old bourgeois neighbourhoods

Scale	Habitus
National developmentalism	*Gecekondu* as *Mahalle*
Individual self-help building (*Gecekondus*)	Migrant networks of ethnic & religious hometown connections
	Central planning and military-bureaucratic milieu

Figure 2.2 State-space mode II: etatist/national developmentalist.

Hollywood onslaught – the other being *Bollywood* – (Kanzler, 2014, pp. 11–18) increasingly saw Istanbul's *gecekondu* as its muse, and socialist realist novelists like Yaşar Kemal and Orhan Kemal found in slum areas both sources of inspiration and a keen audience (Gülhan, 2017). While the migrant *mahalles* silently organised amidst all the deprivation and where first the socialist, then Islamist, grassroots movements persevered against military interventions of 1980 and 1997, academia and public opinion developed a newspeak of the 'dangerous classes'. The *mahalle* of periphery found a new lease in its symbolic power in the next modal arrangement.

Mode III: Neo-liberal/state-corporate alliance state-spaces

The third state-space modality is still emergent. Its genesis can be dated to the break-up of the national developmentalist agenda when the cash-strapped Süleyman Demirel government declared a neo-liberal economic programme on 24 January 1980, or to the heyday of privatisations of publicly owned oil refineries, telecommunications companies, and banks in the early 2000s. Better still, it can be politically defined as the rupture instigated by the general elections on 3 November 2002 that catapulted AKP into power.

This new modality of state-space nexus is incumbent upon the second, national developmentalist mode, though, under the asphyxiating weight of worldwide seizure of neo-liberal policies. Its territorial ambitions are manifestly global. However, Istanbul is not yet a global city, and will possibly never be one. It has tapped into a huge financial valuation of hitherto untouched lands, primarily, erstwhile *gecekondu* settlements, and secondarily, inner-city neighbourhoods with the heavy presence of public ownership of land.

Lived space/territory	Representations of space/place-making
Deindustrialisation	Gated communities
Highways and satellite cities	Gentrification of old bourgeois neighbourhoods
Massive housing investment in the periphery	TV shows
	Gecekondus to apartment buildings

Scale	Habitus
International and national finance-capital circuits	Islamist conservatism
	'Nativism and nationalism'
Financialisation of housing	The new middle class
	White Turks

Figure 2.3 State-space mode III: neoliberal/state-corporate alliance.

The place produced by this novel mode of state-spaces is heavily impressed by a middle-class ethos. The scale this new Istanbul heralds is internationally de-territorialising for the sake of unswerving faith in the powerful growth machine of fictitious capital sunk in the land. Nothing is modest, the sky is the limit. However, no single capitalistic form of accumulation, not even the petrodollars (or, actually, the petro-liras and gold) of the Gulf countries can sustain this limitless expansion. Finally, the habitus in this new mode of state-space in Istanbul is still connected to the ethnically and religiously motivated *mahalles*, Turkishness, and ubiquity of state power that determines the asymptotes of state and space-making.

New state-spaces and the emergence of unhindered boosterism

The decisive end to the passive government stance towards housing and urban planning came with the passage of the Housing and Urban Development Act of 1984. This act established a Housing and Urban Development Fund (TOKI). The Fund was exempt from effective parliamentary control. The Fund (controlled partly by the Housing and Urban Development Administration) was financed by a share of the taxes on tobacco and alcoholic products, in addition to a notorious cut from every employee's payroll and another standard levy on every citizen's foreign travels. However, the TOKI administration was inefficient and lethargic throughout the 1980s and 1990s; its housing projects were merely oriented towards upper middle classes who could afford up to 80 per cent down payments in an already overpriced housing market. From its establishment to early 2003 – until Erdoğan came to power – the administration constructed around 30,000 housing units, and marginally contributed to the financing of the housing industry.

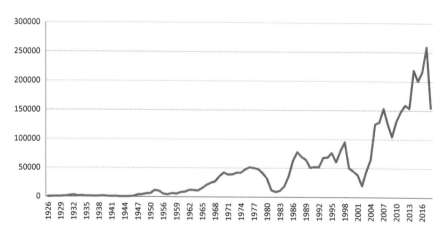

Figure 2.4 Construction permits issued by municipalitites in Istanbul.

Source: Tekeli, 1996, pp. 132–139; Devlet İstatistik Enstitüsü, 1966–1995; TÜİK (TURKSTAT) 2019.

The Erdoğan government was ambitious from the start. Legislative arrangements were made to grant TOKI with the power to borrow from international financial markets without Treasury backing or interference, with the authority of eminent domain for urban renewal projects in *gecekondu* areas, and this extended to the government's full backing regarding the appropriation of public land. In 2003, TOKI embarked upon an unprecedented housing development plan nationwide and was slated to finish 500,000 units by 2013; 100,000 alone were to be built in Istanbul.

In appropriating a free-market logic, TOKI developed all its housing projects through tenders to the private contractors and developers. The process can be more or less summarised as follows: TOKI announces a development project on public land and seeks private parties as development partners. Contractors enter into an agreement with TOKI. In return for development rights on public land, the private corporation allocates a certain portion of the project as subsidised housing to be marketed by TOKI. While the private developers market their apartments at prevailing market prices mostly through mortgages secured by private banks, TOKI has its own financing arm that permitted below market rates and generous terms of credit. TOKI would erroneously label its share in such private developments as social housing. The state assumed the role of erstwhile squatter settlers and fanned the flames of speculation on the built environment by supplying more and more land for the sprawl; meanwhile, TOKI and its contractors expanded the scale to unprecedented levels. That meant a massive inflation in development size. For example, a moderately scaled TOKI housing project would involve the building of 1,400 units. In contrast, a typical *small-scale* apartment building development is limited to 20 to 40 units.

In Istanbul, only a third of TOKI's projects are directed towards lower middle classes, and only two of them were specifically built for economically disadvantaged people. TOKI, KIPTAS, the housing development corporation owned by Istanbul Metropolitan Municipality, and Emlak Konut, another state enterprise

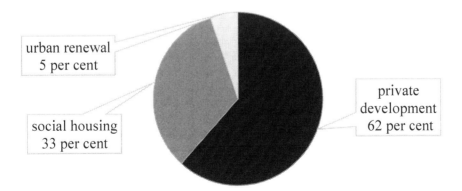

Figure 2.5 State-led housing development projects in Istanbul 2003–2014.
Source: www.toki.gov.tr/illere-gore-projeler.

specialised in urban housing development, the three main actors of state intervention in the built environment are not public housing and urban development administrations in the strict sense of the term; the way they function closely resembles private developers. TOKI and others run after real estate speculation, and they seek a gradual increase of land rents on the territory they control; they purposefully push some developments on hold, while setting others in motion with higher bidders; they deliberately spin some projects as high-rent generating projects, while limiting their already limited social housing developments to the fringes of the city.

In terms of social space, the most decisive development of this third mode was the complete territorial disappearance of the *mahalle*. Istanbul's traditional unit of settlement was obliterated by the rampant speculation in urban land and its traditional urban patterns of a central mosque, public school, and police station – the latter two were the indelible inventions of the previous Kemalist national developmentalism – were squeezed in between high-rise apartment building, or completely obviated by the gated communities.

Yet, while territorially erased from the fabric of the urban landscape, *mahalle* as a juridico-political concept and a phoneme of everyday discourse made a powerful comeback. Recep Tayyip Erdoğan, immediately after his election as president, announced a series of meetings with *mahalle* administrators from all around Turkey. Hitherto unknown and poorly paid figures who are only responsible for keeping residents' registers in their districts, the *mahalle* administrators – *muhtar* in Turkish – were suddenly in the limelight. Their diligence as the progenitors of Turkish and Islamic identity in the rapidly urbanising cities was praised and soon their wages were increased and elections for this lowest level of local administrative positions became once again competitive. *Mahalle* was revived as the solid bedrock of conservative newspeak and although it no longer existed as a physically tangible entity, the nostalgia surrounding it became the keyword for advertisement oriented towards selling new developments. The symbolic power of space was harnessed to develop a new awareness of Turkishness and nativism.

Representations of space were centred on a new middle-class life for the rapidly globalising Istanbul. Especially after the EU accession talks started in 2005 the city was deemed to be the beacon of Turkey's financial and cultural prowess. Almost all publicly-owned banks transferred their headquarters to Istanbul. And once Istanbul was declared as the cultural capital of Europe for 2010 the city's global reputation skyrocketed to unprecedented levels. Industrial development zones began to transform into commercial areas, the historical peninsula and the inner city of Beyoğlu and Kadıköy turned into tourist hubs. Erstwhile upper-class and upper-middle-class areas turned into bohemian staging sites for the so-called creative class while the peripheries were colonised by the ubiquitous gated communities for both conservatives and liberals alike (Candan & Kolluoğlu, 2008; Çavdar, 2016; Geniş, 2007).

This was not the eagerly awaited arrival of the urbanism as a way of life. In the late 1980s and early 1990s, the old bourgeoisie started a typical flight to the suburbs. Akin to the American 'white flight' of the 1970s – minus the *hyperghettoisation* (Wacquant & Wilson, 1989) – and fed on the 'dangerous classes' rhetoric shaped around the squatter settlements (Erman, 2001; Erman & Eken, 2004) the old bourgeoisie and the so-called creative new middle-class sought for presumed safety, Western comforts, and a shared status identity in the gated communities (Öncü, 1997; Öncü & Weyland, 1997; Geniş, 2007; Candan & Kolluoğlu, 2008). However, by the early 2000s, this turned into a curious polarisation of settlement patterns: coastal areas and newly gentrified inner-city neighbourhoods like Beyoğlu, Cihangir, Beşiktaş, and Kadıköy had become the bastion of a secular and liberally oriented population who were conscious of their lifestyle politics. They were seen as the future of the city and variously defined as the 'new' middle-class: highly educated, globally interconnected, talented creative people bent on pluralistic democratic habits, tolerant of the other and not dedicated to the older style of making politics.

This rhetoric of the 'new' middle-class never actually delivered its potential. The so-called 'new' middle-class barely represented a fringe status in Istanbul's local politics. The new 'new' middle-class was simply located elsewhere, in the area limited by the two highways, where more than 80 per cent of Istanbul's population lived. They were erstwhile *gecekondu* owners, the working masses of the periphery. This actual 'new' middle-class was seriously underestimated. They were called by the right-wing politicians and journalists of the early 2000s as the silent majority (Özkök, 2003) in an apparent allusion to American Republican politics (Lassiter, 2013). They are no longer silent; they are no longer the docile *gecekondu* dwellers who were intent to keep encroaching on the urban public lands. Once peripheral *gecekondu* areas turned into the new centres of real estate boom and erstwhile squatters became *nouveaux riches*. The foundations of the *gecekondu* owners' economic power laid in the built environment and their newfound role as urban rentiers.

The habitus, however, had a dualistic existence and this made itself felt on the political field. The old Istanbulites, the ones who lived in the coastal areas were pointedly referred by the power elites of the ascendant conservatism as the 'white Turks'. The officially sanctioned habitus sought for 'nativism and nationalism' (*yerli ve milli*) in defining its core qualities. This new middle-class identity has four main pillars: first, a ubiquity of emphasis on Turkish national identity and an ascendant nationalism permeates all layers of public discourse. Second, this discourse is built around family values (Yılmaz, 2015) ensconced in the historical hagiography of the *mahalle*. Third, observance of religious practices and respect for the religious establishment is requested, but this does not necessarily mean that strict participation from all echelons of society is expected. And fourth, perhaps the most important of them all, necessitates an unswerving allegiance to a powerful centralised Turkish state.

Through the active reconstruction of the *state-space* in Istanbul, *mahalle* had the most eventful transformation. The *mahalle* has had three different meanings under three different modes: first, it began as a physical territory, as the Ottoman Empire's juridico-political delineation of residence based on faith. Second, that

Ottoman term for residential segregation created a sense of belonging – a representation of space akin to where they hailed from – for the migrants from the 1940s on up until the 1980s. The newcomers established their own social space in the slum *mahalles*. And third, with the physical disappearance of *mahalle* in the last three decades, it turned into a *phoneme* in everyday political language that determined us – the 'true Turks and Muslims' – vs. them – the *white Turks* – distinction. That discursive element of the political habitus found its starkest expression in the well-worn cliché that the ones from *mahalle* are the real backbone of Turkish identity, while the so-called *white Turks* who enjoy a luxurious lifestyle in their Western-style homes along the Bosphorus are undeservedly, but vocally, criticising the government based on Western notions – like pluralism, democracy, feminism, social rights, etc. – that are completely alien to the Turkish identity.

In skilfully forging this doxa, Recep Tayyip Erdoğan's government has achieved a twofold feat. The state helped capital to concentrate more and more in every nook and cranny of activities related to the built environment. And in return, the central government excelled in the massive redistribution and financialisation of urban resources – land, planning regulations, collective resources such as water, sewer systems, health, education, and so forth – and instituted relatively egalitarian ownership of property patterns have fed unto a shared illusion of getting richer. If anything, under this new mode, the habitus of the erstwhile migrants, current proud Istanbulites, is augmented.

Conclusion

In the last two decades, homes in Istanbul turned into life savings for the middle-class, making it possible for homeowners to continue spending on the basis of a presumed increase of real estate prices. This is extremely similar to what happened in the core economies of the world system before the 2008–2009 crisis. The third-cut crises as David Harvey (1982, p. 425) described, or the crises rooted in secondary circuit of capital as Lefebvre (2003, pp. 159–163) explained in terms of expanded reproduction of capital have their bases in the built environment and require a reordering of the relationship between territory/practical spaces, places/representations of spaces, scale, and habitus.

The swing between the commodification of urban sphere and the claim to the right to the city is ascendant at these pivotal moments of transformation. In 2018, Istanbul's housing market entered a new phase. Housing construction permits decreased from an all-time high of almost 260,000 units in 2017 to 155,000 units in 2018 (TÜİK (TURKSTAT), 2019). The real estate speculation and urban land rent derived from thence came to a screeching halt with this unprecedented drop of 40 per cent in a single year. This is not merely about the short-sightedness of the decision-makers' short-run irrationality. If we followed Lefebvre and Harvey's capitalist crises theorisations the answer would be evident that internal contradictions of accumulation processes and reproduction of labour power would be inherently solved through the state's central role in reconfiguring the geographical and regional dispersal of capital.

Neil Brenner, on the other hand, argued that inasmuch as theoretical practice is concretely actualised through knowledge effects, the plethora of state projects have produced their own state effects (Brenner, 2004, p. 85). The visible and disguised procedures, interventions, retreats, and gradual encroachment of the *gecekondus*, state contractors, and private developers have been all but carriers of such state projects. In whatever form they have taken, they were the key surface phenomena of the state's activities in space. Still, this is no one-way street. The same is true for the state. In making the spaces of production and reproduction of social relations in Istanbul, the Turkish state has, in turn, succumbed to its own dialectical loophole: *gecekondus* was a brilliant and unprecedented redistribution of wealth, though the irreverent plebeian land grab had unmistakably reshaped the political sphere and made local governments mere replicas of the national, centralised, and monolithic power located in Ankara.

The Turkish state, like any other state, functioned through modal arrangements among four dimensions. These four dimensions roughly equal to, at the territorial level, the mutations of Turkishness from an ethnically based identity that claims suzerainty over an ambiguous contingence that mythically extends from Asia to the Adriatic, and the indivisible unity of the state and the people at home to one that portends a deeper understanding of constitutional citizenship, civil rights, and the multifarious ethnic, cultural, religious identities. The representations of space, in this context, can be best grasped through a not-so-subtle metonymy that has long prevailed in Istanbul, the doxa that invokes ruling by naming – naming and renaming streets, buildings, squares, even whole neighbourhoods – in addition to the state effects that fed into the political machinery. The scale or the constant rescaling of the Turkish state in the last two decades, referred to here, mainly pointed to the tension between a centripetal, a top-to-bottom organised state apparatus and the incipient requests of an internationally integrated (to the global capital circuits and decision-making bodies) and a centrifugally re-territorialised apparatus that is horizontally structured to reformulate a statehood which once aimed at being a full member of the European Union. Currently, there is no question that the centripetal forces have the upper hand. And finally, the habitus itself is schizoid since it represented the means of resilience and transformative potentials of an economic formation that carries the utmost crisis tendencies for the transformation from a small-scale employer-dominated system to one that faintly resembles monopoly capitalism in new clothes.

The logical outcome of almost four decades of sustained neo-liberal policymaking is nothing but populism. We came to reckon that not only in Istanbul but throughout the world, populism is ascendant. A point of bifurcation is in the making. In Istanbul, the changes bound to happen will necessarily follow the asymptotical arrangements determined by the preceding modes of production of space. However, though definitely not warranted, it is still possible to talk about a new state-space modality that puts the emphasis on the right to the city.

References

Allen, W. (1984). The Abdul Hamid II collection. *History of Photography*, *8*(2), 119–145.
Aslan, Ş. (2004). *1 Mayıs Mahallesi 1980 Öncesi Toplumsal Mücadeleler ve Kent*. Istanbul: İletişim.
Ayata, S. (2008). Migrants and changing urban periphery: Social relations, cultural diversity and the public space in Istanbul's new neighbourhoods. *International Migration*, *46*(3), 27–64.
Ball, M. (1981). The development of capitalism in housing provision. *International Journal of Urban and Regional Research*, *5*(2), 145–177.
Batuman, B. (2006). Turkish urban professionals and the politics of housing, 1960–1980. *METU JFA*, *23*(2), 59–81.
Bayat, A. (1997). *Street politics: Poor people's movements in Iran*. New York, NY: Columbia University Press.
Bilsel, C. (2011). 'Les transformations d'Istanbul': Henri Prost's planning of Istanbul (1936–1951). *A| Z ITU Journal of the Faculty of Architecture*, *8*(1), 100–116.
Boratav, K. (1981). *Tarımsal yapılar ve kapitalizm*. Ankara: Ankara Üniversitesi Basımevi.
Boratav, K., Keyder, Ç., & Pamuk, Ş. (1984). *Kriz, Gelir Dağılımı ve Türkiye'nin Alternatif Sorunu*. İstanbul: Kaynak Yayınları.
Bourdieu, P. (1970). The Berber house or the world reversed. *Information (International Social Science Council)*, *9*(2), 151–170.
Bourdieu, P. (1977). *Outline of a theory of practice*. Cambridge: Cambridge University Press.
Bourdieu, P. (1984). Espace social et genèse des 'classes'. *Actes de la Recherche en Sciences Sociales*, *52*(1), 3–14.
Bourdieu, P., & Wacquant, L. (2001). Neoliberal newspeak: notes on the new planetary vulgate. *Radical Philosophy*, *105*. Retrieved from www.radicalphilosophy.com/commentary/newliberalspeak.
Bozdoğan, S. (2001). *Modernism and nation building: Turkish architectural culture in the early republic*. Seattle, WA: University of Washington Press.
Brenner, N. (1997). State territorial restructuring and the production of spatial scale: Urban and regional planning in the Federal Republic of Germany, 1960–1990. *Political Geography*, *16*(4), 273–306.
Brenner, N. (1999a). Beyond state centrism? Space, territoriality, and geographical scale in globalization studies. *Theory and Society*, *28*(1), 39–78.
Brenner, N. (1999b). Globalisation as Reterritorialisation: The re-scaling of urban governance in the European Union. *Urban Studies*, *36*(3), 431–451.
Brenner, N. (2001). The limits to scale? Methodological reflections on scalar structuration. *Progress in Human Geography*, *25*(4), 591–614.
Brenner, N. (2004). *New state spaces: Urban governance and the rescaling of statehood*. Oxford: Oxford University Press.
Brenner, N., Peck, J., & Theodore, N. (2010). Variegated neoliberalization: Geographies, modalities, pathways. *Global Networks: A Journal of Transnational Affairs*, *10*(2), 182–222.
Candan, A. B., & Kolluoğlu, B. (2008). Emerging spaces of neoliberalism: A gated town and a public housing project in Istanbul. *New Perspectives on Turkey*, *39*, 5–46.
Castells, M. (1976a). Is there an urban sociology? In C. G. Pickvance (Ed.), *Urban sociology: Critical essays* (pp. 33–59). New York, NY: St. Martin's Press.
Castells, M. (1976b). Theory and ideology in urban sociology. In C. G. Pickvance (Ed.) *Urban sociology: Critical essays* (pp. 60–84). New York, NY: St. Martin's Press.

Castells, M. (1977). *The urban question : A Marxist approach*. London: Edward Arnold.
Castells, M. (1978). *City, class, and power*. London: Macmillan.
Çavdar, A. (2016). Building, marketing and living in an Islamic gated community: Novel configurations of class and religion in Istanbul. *International Journal of Urban & Regional Research, 40*(3), 507–523.
Checkoway, B. (1980). Large builders, federal housing programmes, and postwar suburbanization. *International Journal of Urban and Regional Research, 4*(1), 21–45.
Dagnaud, M. (1983). A history of planning in the Paris region. *International Journal of Urban and Regional Research, 7*(2), 219–236.
Erman, T. (2001). The politics of squatter (Gecekondu) studies in Turkey: The changing representations of rural migrants in the academic discourse. *Urban Studies, 38*(7), 983–1002.
Erman, T., & Eken, A. (2004). The 'other of the other' and 'unregulated territories' in the urban periphery: Gecekondu violence in the 2000s with a focus on the Esenler case, Istanbul. *Cities, 21*(1), 57–68.
Geniş, Ş. (2007). Producing elite localities: the rise of gated communities in Istanbul. *Urban Studies, 44*(4), 771–798.
Gottdiener, M. (1985). *The social production of urban space*. Austin, TX: University of Texas Press.
Goursolas, J. M., & Atlas, M. (1980). New towns in the Paris metropolitan area: An analytic survey of the experience, 1965–79. *International Journal of Urban and Regional Research, 4*(3), 405–421.
Gül, M. (2009). *The emergence of modern Istanbul*. London and New York, NY: I. B. Tauris.
Gül, M. M., & Lamb, R. (2004). Urban planning in Istanbul in the early republican period: Henri Prost's role in tensions among beautification, modernisation and peasantist ideology. *Architectural Theory Review, 9*(1), 59–81.
Gülhan, S. T. (2017). Ezilmiş ve Aşağılanmışlar: 1960'lar Türkiye'sinde Gecekondu Meselesi. *Mülkiye Dergisi, 41*(1), 195–230.
Gürsoy, Ö. (1970). Tek bir ses vardı: Ölüyoruz, ölüyoruz! *Milliyet*, 18 October, pp. 1, 9.
Harvey, D. (1982). *The limits to capital*. Oxford: Basil Blackwell.
Harvey, D. (2005). *A brief history of neoliberalism*. Oxford and New York, NY: Oxford University Press.
Hawley, A. H. (1944). Ecology and human ecology. *Social Forces, 22*(4), 398–405.
Hopkins, T. K., Wallerstein, I., Kasaba, R., Martin, W. G., & Phillips, P. D. (1987). Incorporation into the world-economy: How the world-system expands. *Review (Fernand Braudel Center), 10*(5/6), 761–902.
İnalcik, H. (1990). Istanbul: An Islamic city. *Journal of Islamic Studies, 1*(1), 1–23.
İnalcık, H. (2015). *Tarihe Düşülen Notlar Cilt II*. İstanbul: Timaş Yayınları.
İpekçi, A. (26 October 1970). Her Hafta Bir Sohbet: Kolera'yı Nasıl Atlattım. *Milliyet*, p. 9.
Jessop, B., Brenner, N., & Jones, M. (2008). Theorizing sociospatial relations. *Environment and Planning. D, Society and Space, 26*(3), 389–401.
Kanzler, M. (2014). *The Turkish film industry: Key developments 2004 to 2013*. Strasbourg: Observatoire Européen de L'Audiovisuel (Council of Europe).
Keleş, R. (2010). *Kentleşme Politikası* (11. Baskı). Ankara: İmge Kitabevi.
Keyder, Ç. (1976). *Emperyalizm azgelişmişlik ve Türkiye*. İstanbul: Birikim Yayınları.
Keyder, Ç. (1983). Small peasant ownership in Turkey: Historical formation and present structure. *Review (Fernand Braudel Center), 7*(1), 53–107.
Keyder, Ç. (1987). *State and class in Turkey : A study in capitalist development*. London and New York, NY: Verso.

Kuban, D. (2000). *İstanbul, bir kent tarihi: Bizantion, Konstantinopolis, İstanbul.* İstanbul: Türkiye Ekonomik ve Toplumsal Tarih Vakfı.

Lassiter, M. D. (2013). *The silent majority: Suburban politics in the sunbelt south.* Princeton, NJ: Princeton University Press.

Lefebvre, H. (1991). *The production of space.* Oxford, UK and Cambridge, MA: Blackwell.

Lefebvre, H. (2003). *The urban revolution.* Minneapolis, MN: University of Minnesota Press.

Lojkine, J. (1976). Contribution to a Marxist theory of capitalist urbanization. In C. G. Pickvance (Ed.), *Urban sociology: Critical essays* (pp. 119–146). New York, NY: St. Martin's Press.

Lojkine, J. (1977). L'analyse marxiste de l'état. *International Journal of Urban and Regional Research, 1*(1), 19–23.

Mardin, Ş. (1971). Ideology and religion in the Turkish revolution. *International Journal of Middle East Studies, 2*(3), 197–211.

Mardin, Ş. (1973). Center-periphery relations: A key to Turkish politics? *Daedalus, 102*(1), 169–190.

Mardin, Ş. (1995). *Türk modernleşmesi* (4. Baskı). İstanbul: İletişim.

Marston, S. A. (2000). The social construction of scale. *Progress in Human Geography, 24*(2), 219–242.

Marston, S. A., & Smith, N. (2001). States, scales and households: limits to scale thinking? A response to Brenner. *Progress in Human Geography, 25*(4), 615–619.

Öncü, A. (1997). The myth of the 'ideal home' travels across cultural borders to Istanbul. In Öncü, A. & Weyland, P. (Eds.), *Space, culture and power: New identities in globalizing cities* (pp. 56–72). New York, NY: Zed Books.

Öncü, A., & Weyland, P. (1997). *Space, culture and power: New identities in globalizing cities.* New York, NY: Zed Books.

Özbek, M. (2013). *Popüler Kültür ve Orhan Gencebay* (11. Baskı). İstanbul: İletişim Yayınları.

Özkök, E. (7 June 2003). Dili Çözülmüş Sessiz Çoğunluk. *Hürriyet,* p. 12.

Pahl, R. E. (1977). Stratification, the relation between states and urban and regional development. *International Journal of Urban and Regional Research, 1*(2), 7–18.

Park, R., & Burgess, E. (1984). *The city.* Chicago, IL: University of Chicago Press.

Préteceille, E. (1981). Collective consumption, the state, and the crisis of capitalist society. In M. Harloe and E. Lebas (Eds.), *City, class, and capital : New developments in the political economy of cities and regions* (pp. 1–16). London: Edward Arnold.

Sargın, G. A. (Ed.) (2002). *Ankara'nın kamusal yüzleri: başkent üzerine mekân-politik tezler.* İstanbul: İletişim.

Saunders, P. (1981). *Social theory and the urban question.* London: Hutchinson.

Shaw, W. M. K. (2009). Ottoman photography of the late nineteenth century: An 'innocent' modernism? *History of Photography, 33*(1), 80–93.

Soja, E. W. (1980). The socio-spatial dialectic. *Annals of the Association of American Geographers, 70*(2), 207–225.

Soja, E. W. (1989). *Postmodern geographies: The reassertion of space in critical social theory.* London and New York, NY: Verso.

Soja, E. W. & Hadjimichalis, C. (1979). Between geographical materialism and spatial fetishism: Some observations on the development of Marxist spatial analysis. *Antipode, 11*(3), 3–11.

Stokes, M. (1998). *Türkiye'de Arabesk Olayı.* İstanbul: İletişim Yayınları.

Tekeli, İ. (1994). *Development of the Istanbul metropolitan area: urban administration and planning.* İstanbul: IULA.

Therborn, G. (2017). *Cities of power: The urban, the national, the popular, the global*. London: Verso.
TÜİK (TURKSTAT). (2019). Yapı İzin İstatistikleri (Database). Retrieved under https://biruni.tuik.gov.tr/yapiizin/giris.zul.
Wacquant, L. (2018). Bourdieu comes to town: Pertinence, principles, applications. *International Journal of Urban & Regional Research, 42*(1), 90–105.
Wacquant, L., & Wilson, W. J. (1989). The cost of racial and class exclusion in the inner city. *The Annals of the American Academy of Political and Social Science, 501*(1), 8–25.
Wallerstein, I. (1974). *The modern world-system I: Capitalist agriculture and the origins of the European world-economy in the sixteenth century*. Cambridge: Academic Press.
Wallerstein, I., Decdeli, H., & Kasaba, R. (1987). The incorporation of the Ottoman Empire into the world-economy. In H. İslamoğlu-İnan (Ed.), *The Ottoman Empire and the world-economy* (pp. 88–97). New York, NY: Cambridge University Press.
Yılmaz, Z. (2015). 'Strengthening the family' policies in Turkey: Managing the social question and armoring conservative-neoliberal populism. *Turkish Studies, 16*(3), 371–390.

3 The city as a business

Nicole Ruchlak and Carsten Lenz

Introduction

'*Stadtluft macht frei*' – city air makes you free. This German saying is about 1,000 years old, and it describes a principle of medieval law which allowed bondsmen and bondswomen who moved to the city to gain their freedom on condition that they stayed behind the city wall for a year and a day.[1]

Today, some ambitious visionaries are proposing a new link between the city and freedom. As in medieval times, in their visionary view the city has the power to set people free. However, the freedom they have in mind is a different one than medieval law applied to, and one that existing cities cannot provide because they are part of existing states and controlled by them – which, they argue, is the very opposite of freedom. They therefore propose that new cities are built. These cities are meant to be independent from political and state control – to the visionaries' thinking, the only motor needed to drive them and citizenship is free-market economy and competition.

This idea is the basis of some concepts we would like to present and analyse here. The concepts are part of a liberalist economic logic that has developed over the last few decades. Up to now, proponents of this logic have tried to implement their ideas within the existing political order. Now, there are more radical ideas being formulated and pursued that deal with getting rid of the existing political order altogether. This, at least, becomes clear from the following concepts. As we will show, these urban-state visions not only reflect global trends; they also push them forward. Their proponents aim at eliminating democratic structures, arguing that this is for the good of all. They want to abandon the essentially political concept of citizenship and transform the city's inhabitants into customers not merely by presenting their models, but by establishing a worldwide network and putting their ideas into practice. In this chapter we elaborate this observation by presenting and analysing examples of several concepts.

Autonomous states – Charter Cities and ZEDE

In 2009, the renowned US economist and later Nobel laureate Paul Romer gave a TED talk on how poor countries could improve their economies (Romer, 2009). His idea, the so-called Charter Cities, seems quite simple: developing countries should designate certain uninhabited or sparsely populated areas of their territory to build and create new prosperous and safe cities. These new Charter Cities would be exempt from the regulations of the state – they would have their own laws and their own administration. Instead of being controlled by the state, they would be governed by a body of external experts. Ideally, these experts would come from developed Western democracies. Their expertise and mandate are supposed to provide legal certainty, which in turn should attract investors.

Even if this outline suggests that the Charter Cities are nothing more than just another version of the special economic zones that exist around the world, with their reduced taxes and tariffs, the concept of Charter Cities is far more ambitious. Instead of being part of the state, while enjoying some modified exceptions from taxes, they would be disconnected from it, forming a state within a state, based on and operating by their own rules. Paul Romer's concept rests on his observation of the desolate situation of many poor countries, the main problem of which is supposed to be their 'bad rules'. Logically, in his view, the obvious remedy is to introduce 'good rules'. But since many of these countries seem unable to implement 'good rules' themselves, he proposes that the task should fall to external experts who govern a specially designated state-free territory (Romer, 2010).

Romer's vision of a state within a state seems to be a good fit for a country like Honduras, which is very poor and has an extremely unequal distribution of income and a very high murder rate (77.7 homicides per 100,000 inhabitants in 2010).[2] That at least seemed to be Porfirio Lobo's opinion, when he decided to put Romer's theory into practice. In 2010, the then Honduran President and his chief of staff, Octavio Sánchez, took the first steps to establish Charter Cities in Honduras.

After years of debate, several draft bills and amendments to the constitution, a law was finally passed in 2013 that proposed details of a '*Zona de Empleo y de Desarollo Económico*' – Zone of Employment and Economic Development – or ZEDE for short (Government of Honduras, 2013). With the project still in its infancy, it is not yet entirely clear which parts of the country should become designated ZEDEs, but the government, under the current president Juan Orlando Hernandez, frequently talks about some regions and about international investors who are supposedly interested.[3]

Even though the concept has not yet been realised, there are clear guidelines and ideas for these future Charter Cities: in a country driven by high crime rates, it goes without saying that security for ZEDE investors would be of paramount importance. It is therefore stipulated that these zones would have their own police force, separate from the state police. The Honduran constitution would apply only to a few aspects: foreign policy, defence, Honduran citizenship matters, and some parts of criminal law. Everything else is decided completely independently from the Honduran state, as it is seen to be exclusively a matter of the sovereignty of the experts who govern the ZEDE. This also includes the inhabitants of the ZEDE not having the right to participate in decision-making.[4] Their role is limited to working in the city-state, and they are not supposed to act as citizens. They are not meant to have a voice in the politics of the ZEDE, as democratic participation is eliminated by ZEDE law. Instead, rule is handed over to 'CAMP', the main governing body of the ZEDE. CAMP is an acronym for *Comité para la Adopción de Mejores Prácticas*, or in English: Committee for the Adoption of Best Practices. CAMP is not an elected body but one that is installed by the Honduran government. The law does not mention any mechanism for how CAMP could be held responsible for its decisions. This is quite remarkable given its extensive powers to create the laws and rules for the zone, nominate judges, and appoint those in charge of the zone's administration (Government of Honduras, 2013, Art. 11).

This reflects Paul Romer's initial idea of a government by experts, and it raises an important question: What exactly is an expert? The Honduran government gave a clear answer to this question when it proposed the following persons in 2013: four previous members of the Reagan Administration who are fervent proponents of libertarian ideals and free-market economy; the president of the Hayek Institute in Vienna, Barbara Kolm, who later was candidate as a minister of economic affairs for the right-wing Freedom Party of Austria (FPÖ); and three members of the Cato Institute, which is committed to 'individual liberty, limited government, free markets, and peace' (Government of Honduras, 2014). The overall ideological orientation of CAMP is probably best reflected in a statement by another member, Grover Norquist, an influential US anti-tax activist: 'I don't want to abolish government. I simply want to reduce it to the size where I can drag it into the bathroom and drown it in the bathtub' (Liasson, 2001). Meanwhile, CAMP has been reduced from 21 to 12 members, with five foreign experts (Economist, 2017, p. 35). Details are hard to come by, since the Honduran government does not provide specific information.

As to the future socio-economic development of the ZEDE, the autonomous city-state, proponents appear confident that high-quality jobs would be created in the zone through maximum freedom from regulation. Such deregulation appeals to people like Michael Strong, an entrepreneur who has followed the developments in Honduras closely from the beginning and has spoken with those responsible. He is convinced that the pharmaceutical industry will profit from deregulation: whereas the pharmaceutical industry is hampered by strict regulation in countries like the US, in the ZEDE the long process of licensing drugs could be shortened considerably (Ruchlak, 2014). There are similar ideas in other sectors, but none has yet been presented in detail. The same applies to information about the basic infrastructure that makes a city habitable, such as social services, health system, pensions, unemployment insurance, schools, and hospitals. The ZEDE programme also fails to clarify who would be responsible for building streets and houses or for providing public transportation.

Barbara Kolm, chairperson of the CAMP board, is however confident in the mechanism of the free market to solve these problems. If there is demand for certain services, she argues, then there will be companies to provide them. If there is no company offering special services, then no such services are needed. Otherwise, the market would have taken care of them. Like other proponents of

ZEDEs, Kolm believes in the power of private companies and in the power of competition. The notions of freedom and competition shape the ideological core of their motivation. This ideology of free and unlimited competition is meant to be realised in the ZEDE. As Kolm puts it: 'The ZEDE allows companies to work under ideal competition and market conditions'. If the project is a success, she says, 'we rewrite economic history' (Ruchlak, 2015).

Democracy is missing from this concept, but this is no shortcoming according to the ZEDE visionaries. They argue that while residents have no influence on the organisation of such a model city, they are always free to leave if they do not like it. In this way, the new ZEDE city-state purports to offer something that is missing anywhere else: real individual freedom.

In terms of ideology, the project is based on – an often justified – mistrust of governments in states like Honduras. (Yet, at the same time, for its realisation it relies on the very same government). Associated with this mistrust is a fundamental scepticism towards processes of democratic political participation, rooted in two ideas. First, a very old liberal idea portrays democracy as a threat to individual freedom, especially to the right of property. The second idea that motivates the visionary concepts described here, claims that there is an alternative to the political processes that hitherto governed existing societies. This alternative mechanism is the free market, which is thought to be not only much more efficient in catering for all individual needs, it also makes it possible to abandon the search for compromise and the political battles with their unforeseeable results.

Even though this concept means a threat to the democratic process, there may be one argument in its favour: Why would it be wrong to give the experiment a try? There is a straightforward answer. It would be wrong because the experiment is badly designed. It is based on the idea of 'good rules' being introduced in a country that is dominated by bad rules. Good rules are a key factor for a country's prosperity, as Paul Romer stated.[5] This is a reasonable point to make. But one important condition must be met: democracy should be part of any set of good rules. For democracy is an important method of improving rules and a mechanism to prevent good rules being distorted or applied only in favour of powerful interest groups. The way ZEDE is being realised in Honduras shows that these concerns are justified.

Free Private Cities

Whereas ZEDE while having an autonomous status is still embedded in an existing state, other concepts attempt to abolish the state completely and establish a free-market society without any political regulation at all. One example is the idea of so-called Free Private Cities as proposed by the German entrepreneur Titus Gebel. Together with an international team, he revealed his vision in a series of newspaper articles, conferences, a book, and on his website (Gebel, 2018; Free Private Cities, 2019).

Free Private Cities are essentially a business model based on the idea of cities as a commodity offered to customers. At first glance, these newly built cities appear to be similar to privately run gated communities that aim to offer security and facilities to their residents. However, Free Private Cities are meant to achieve much more ambitious political objectives by confronting existing states with the model of a radically different community. Designed to appeal to people who feel severely restricted in existing societies, Free Private Cities offer a stateless, privately run alternative.

Titus Gebel and his team envisage that a company – the operator of the Free Private City – provides the functions of a city that would normally be taken care of by the state or municipality: protection of life, freedom, and property. To receive these services the inhabitants pay a regular fee. Payment and details of the services provided in return are the subject of a so-called *citizen's contract*, which each resident would sign. This contract specifies all duties of the citizens and the operator. If additional services are desired, the fee increases. In contrast to states with a general tax system, the principle that applies in a Free Private City is the same as in any other private contract: you get what you pay for, and you pay for what you get (Gebel, 2018, pp. 101–104).

If contracts are violated, both parties can resort to independent arbitration like in international commercial law. However, Titus Gebel argues, this will only happen in exceptional cases, because the market forces will have two beneficial effects: first, homogeneity of the city's population; second, services that exactly match the residents' requirements. The latter, he claims, are ensured by the fact that the operator tries to attract paying citizens to make profits and, ideally, competes with other operators (Gebel, 2018, pp. 111–113).

Since operators are free to admit or refuse potential inhabitants, Free Private Cities are a kind of private members club, and citizenship becomes a sort of club good. One possible criterion for accepting citizens could be religion; another could be wealth. Correctly applied, these criteria would lead to homogeneous and hence harmonious societies, Titus Gebel argues. According to his understanding of society, this elimination of social conflicts ultimately makes political debates, controversy, and bargaining redundant. Instead of democratic participation, which gives members of society the possibility to seek a compromise between different interests, the Free Private City provides the option to cancel the citizen contract and leave the city. In Titus Gebel's view, this leads to a homogeneity that removes the need for democratic participation in general. And this indeed is the explicit goal of this concept of Free Private Cities: democracy, as a rule by the majority, permanently threatens individual freedom and property and should, therefore, be abolished (Gebel, 2018, pp. 35, 92–93). In return, the members of the club enjoy freedom in the sense of self-determination in a voluntary association.

The concept of Free Private Cities is more radical than that of the ZEDE. In the latter, citizens are reduced to a labour force, whereas inhabitants of the Free Private City are seen as customers enjoying the city as a commodity. Both concepts leave no room for citizenship in the political sense of democratic participation. Instead, economics is meant to replace politics. This radical market ideology has its adherents. At a conference at the University of Zurich in 2017,

Gebel's concept was presented as a possible solution to problems originating from migration (Aerni, 2017); Germany's main employers' association has given Gebel space on its homepage to describe his ideas in detail (Gebel, 2017); and influential newspapers such as the German weekly *Wirtschaftswoche* (Fischer & Guldner, 2017) and Swiss quality paper *Neue Zürcher Zeitung* (Fuster, 2017) have published a number of articles sympathetic to his vision.

Additionally, Gebel and his followers are part of an international network of like-minded organisations. One of them is the Seasteading Institute, which shares goals similar to the concept of Free Private Cities. It should therefore come as no surprise that Titus Gebel is a member of its board of trustees (Seasteading Institute, 2019a).

Seasteading

Seasteading is an even more radical concept for building new societies than ZEDE and Free Private Cities. It is promoted by the Seasteading Institute, which was originally founded to 'test new ideas for government' (Seasteading Institute, 2019b). Randolph Hencken, the former managing director of the Seasteading Institute, is hands-on: he regards democracy as an 'outdated technology'; even though it has brought wealth, health, and happiness to billions of people, it is time to try something completely new, he says (Gaertner, 2014). Since all the land is already ruled by governments, only the oceans are free for conducting this kind of experiment.

The idea behind seasteading is to build human settlements on large floating platforms on the open sea, at least 12 nautical miles off the coast, and beyond the jurisdiction of existing states. The advantage of seasteading is that companies could operate and people could live in the settlements without any regulation by the state. This is how the founders of the Seasteading Institute define the kind of freedom they want to promote. One example of the beneficial effects of this unrestricted freedom, frequently cited by its proponents, is the opportunity to do medical research without governmental limitations (Seasteading Institute, 2019c; Quirk & Friedman, 2017, pp. 241–246). The same would apply to IT specialists, who could work untroubled by data protection laws. This deregulated environment of seasteading appeals to people like Larry Page, former Google CEO:

> I think as technologists we should have some safe places where we can try out some new things and figure out what is the effect on society, what's the effect on people, without having to deploy kind of into the normal world.
> (Yarow, 2013)

More recently, the seasteading movement has scaled its concept down, moving it from the open sea and closer to the coast. The reason is a purely practical one, it seems: building the platforms presents numerous technical challenges, and therefore they would need to be built initially within the jurisdiction of a state. Proponents argue that such prototypes provide experience for constructing the real thing further off-shore at a later stage (Quirk & Friedman, 2017, p. 59). In addition, the institute is seemingly trying to broaden its follower base by appealing to wider concerns of health, environmental protection, and reducing poverty. The official website and the latest book to propagate the idea of seasteading are both dominated by representations of practical ideas to generate clean energy, sustainably grow food, or build floating houses on the ocean (Quirk & Friedman, 2017; Seasteading Institute, 2019c). This shift of emphasis has already caused some irritation among supporters of the liberalistic cause (Doherty, 2017).

In 2017, the Seasteading Institute signed an agreement with the government of French Polynesia to build a platform off the coast of one of its islands (Gelles, 2017). A glossy high production value video by Blue Frontiers – a company founded basically by the Seasteading Institute to realise seasteading projects – shows a cluster of beautifully and sustainably designed floating houses, anchored to a central platform (Blue Frontiers, 2019). These floating houses are the physical basis of one of the central ideas that the seasteading movement shares with Free Private Cities: state-like entities offering their services and competing with each other for residential customers. As citizens can react to decisions of a city's management simply by moving to another territory, governing administrations could be held accountable for their actions much more powerfully than is possible in existing states – an idea already put forward in a thought experiment in the important libertarian manifesto *The Machinery of Freedom* by David Friedman ([1973] 1989, p. 123).

Patri Friedman, David Friedman's son and co-founder of the Seasteading Institute, calls it 'voting with your house' (Friedman & Taylor, 2012, p. 224): those who dislike one seastead only have to travel to another with their floating house. According to this concept, the politically relevant freedom of the inhabitants consists of their freedom of choice as consumers of government services. The means of influencing the governing bodies of these cities or states is what Albert O. Hirschman called the 'exit' option, as opposed to the essentially political option of voicing discontent and trying to change a society from within (Hirschman, 1970, p. 15). The option to change seasteading society from within by democratic participation is not an element that is put forward in the concept. The only aspect that is emphasised is individual freedom; and this individual freedom comes down to the principle of make or break – in essence, if you do not like it, leave. That is exactly the same principle as the Free Private City and ZEDE concepts are based on, although in the on-shore cities the cost of moving out is supposedly higher than in the visions of floating cities on the ocean.

On the way to implementing these cities, however, seasteading is confronted with the problems of political reality. At the beginning of 2018, the Polynesian government distanced itself from the seasteading plans after some political controversies. It is not entirely clear to what degree the decision was influenced by the public opposition to the project. At any rate, part of the opposition was motivated by the suspicion that the aim of seasteading was simply to create a tax haven (Robinson, 2018). For the moment, seasteading, as well as Free Private Cities and ZEDE, is essentially an ambitious idea of some 'visionaries' in search of investors and sites. Although in Honduras the conditions for realising ZEDEs are already in place, none of the three concepts for founding a new (city-)state has been put into practice so far. Nevertheless, there are at least three important reasons for taking these visions seriously. First, they are symptoms of a decade-long liberalist trend. Second, they re-enforce this trend by radicalising it and attempting to realise their ideas in concrete city-building projects. And third, this is accomplished by an international network of influential people, who meet at conferences, hold powerful positions in political organisations, publish sympathetic depictions of each other's ideas, and invest in each other's projects. It is not surprising, then, that some of these people hold important posts in several of the projects described. Titus Gebel, the founder of the Free Private Cities corporation, is also a member of the board of trustees of the Seasteading Institute and invests in a ZEDE project in Honduras (Gebel, 2018, p. 156). Michael Strong has expressed some interest in ZEDE investment, too, and was a member of the board of trustees of the Seasteading Institute (Strong, 2017).

The visionaries of the concepts presented are not only interconnected with each other, but they also have excellent connections to governments. An example is Michael Klugman, one of the initiators of ZEDE in Honduras and a former member of its governing body CAMP – he introduced Honduran representatives of the ZEDE project to influential members of the Trump administration (Dada, 2017). Peter Thiel, the founder of PayPal, who initially co-founded the Seasteading Institute, was even a member of Donald Trump's presidential transition team (Woolf & Wong, 2016). Barbara Kolm, who now leads the governing body of ZEDE, is a member of the right-wing FPÖ (Freedom Party of Austria), which was part of the government in Austria, and she is vice-president of the General Council of Austria's national bank (Österreichische Nationalbank, 2019). She is also president of the Austrian Hayek Institute, a think tank that promotes libertarian ideas. Probably the most influential of these libertarian think tanks is the Cato Institute. Several of its researchers, fellows, and speakers are connected to ZEDE, the Seasteading Institute and Free Private Cities: Michael Klugman, for example, member of ZEDE's governing body CAMP, and Patri Friedman (founder of the Seasteading Institute).

This network not only allows an international exchange of ideas but also creates a global discourse that corresponds to the global market economy. Increasingly, these visions are being promoted as solutions to complex and large-scale problems, as political geographer Casey Lynch states (Lynch, 2017, p. 90). Even though it is not certain when and if the presented concepts will be implemented, the discourse itself has an impact on society, for it frames the political issues:[6] so far, representatives of the economic libertarian positions have only criticised the state and political order in theory, while now they are proposing alternative models of society beyond the state. Instead of integrating economic ideas into a governmental framework, the governmental framework is abandoned in favour of a new state-like structure that follows radical free-market ideas.

All approaches seek to minimise or eliminate the influence of state actors on society, which includes a fundamental mistrust of democratic participation by citizens. In contrast, there is almost unlimited trust in the free market and its mechanisms as a working model for more or less all aspects of society. It is not the political authorities or the citizens who are meant to make decisions; instead, the decisions are mainly left to the market (Friedman & Taylor, 2012). The ideas presented here are therefore symptomatic of the current tendency to minimise the state's influence by subordinating society to economic logic. This logic contains another aspect that can currently be observed in the nation-state, too: the tendency to seal oneself off from strangers. This statement is true at least as far as the concept of seasteading and Free Private Cities is concerned. Instead of a diversity of interests and pluralism, the city/state models promote a union of like-minded people – for example, with regard to religion, wealth, and political orientation. There is no space for contradictory interests and divergent views that require political process and discussion.

The models presented here have become increasingly widespread in recent years. What is more, their proponents are investing money and effort to put them into practice. And this might not only result in the creation of cities 'somewhere out there' but will also have an effect on society in general, impacting on the reality of citizenship and democracy in a fundamental way.

Notes

1 This research benefitted from a scholarship funded by Max-Planck-Institut für Gesellschaftsforschung (Max Planck Institute for the Study of Societies), Cologne, Germany.
2 According to the World Bank, 60.9 per cent of the population live in poverty by national standards, and around 20 per cent of the population live in extreme poverty, defined by a daily income of less than 1.90 US$ (World Bank, 2018). Over 70 per cent of the landholders own about 8 per cent of agricultural land. More than 60 per cent of agricultural land is owned by 5 per cent of the landholders (Instituto Nacional de Estadística, 2008, p. 15). The homicide rate meanwhile decreased to 43.6 per cent in 2017, but it is still high by international standards (see Igarapé Institute, 2019).
3 One of the latest examples of optimistic but vague promises is La Tribuna, 2018 (La Tribuna, 2018).
4 The same applies to Hondurans who live in an area designated as a ZEDE region. The ZEDE law allows them to be expropriated, if necessary. Even though they are granted financial compensation, there is no information about the amount. The topic of landownership has been a highly conflict-laden matter for centuries. Small farmers and indigenous communities rarely possess official documents proving their ownership. This often results in conflicts between small farmers on the one hand and large landowners and agro-industrial enterprises on the other. The state does not protect those who are threatened – among whom there are first and foremost small farmers (see Edelman & León, 2013).
5 Paul Romer, who inspired the concept of ZEDE, has since voiced his disappointment with how the Charter Cities would be implemented in Honduras and has long since left the project.
6 Steinberg, Nyman and Caraccioli describe this with reference to the Seasteading project:

> as a mechanism that utilises marine romanticism and science fiction fantasy to spur a critique of twentieth-century state-regulated capitalism, the seasteading

movement can be seen as one wedge of a much larger neoliberal project in which the "free medium" of the ocean frequently plays a leading role.

(Steinberg, Nyman, & Caraccioli, 2012, p. 1545)

References

Aerni, P. (2017). Jenseits des Territorial-Asyls. Paper presented at *Zürcher Migrationskonferenz September 7th 2017*, Zurich.

Blue Frontiers (2019). Welcome to Blue Frontiers. *Blue Frontiers*. Retrieved from www.blue-frontiers.com/en/.

Dada, C. (2017). Honduras y su experimento libertario en el golfo de Fonseca. *El Faro*, 20 April. Retrieved from elfaro.net/es/201704/centroamerica/20283/Honduras-y-su-experimento-libertario-en-el-golfo-de-Fonseca.htm.

Doherty, B. (2017). Seasteading in paradise. New promise for floating free communities in a Polynesian lagoon – but is the movement leaving libertarianism behind?. *Reason*, 21 May. Retrieved from reason.com/archives/2017/05/21/seasteading-in-paradise.

Economist (2017). A shadowy experiment. *The Economist*, 12 August, 34–35.

Edelman, M., & León, A. (2013). Cycles of land grabbing in Central America: An argument for history and a case study in the Bajo Ajuán, Honduras. *Third World Quarterly*, *34*(9), 1697–1733.

Fischer, M., & Guldner, J. (2017). Wenn das Leben ohne Staat möglich wird. *Wirtschaftswoche*, 8 December. Retrieved from www.wiwo.de/my/politik/deutschland/privatstaedte-soziale-revolution/20680404-2.html?ticket=ST-2340545-w3XW1hEDfXz6j71BllgP-ap1.

Free Private Cities (2019). Business Model. *Free Private Cities*. Retrieved from freeprivatecities.com/business-model/.

Friedman, D. ([1973]1989). *The machinery of freedom. Guide to radical capitalism*. LaSalle: Open Court Publishing Company.

Friedman, P., & Taylor, B. (2012). Seasteading: Competitive government on the ocean. *Kyklos*, *65*(2), 218–235.

Fuster, T. (2017). Leben ohne staatliche Gängelung. *Neue Zürcher Zeitung*, 25 June. Retrieved from www.nzz.ch/wirtschaft/wig-ld.1302811.

Gaertner, J. (2014). Mikrogesellschaften. Hat die Demokratie ausgedient?. *Capriccio*, 5 May. Bayerischer Rundfunk – Fernsehen, Munich.

Gebel, T. (2017). Von Freien Reichsstädten zu Freien Privatstädten. Vorschlag für eine alternative Ordnung. *Deutscher Arbeitgeberverband*, 13 February. Retrieved from https://deutscherarbeitgeberverband.de/aktuelles/2017/2017_02_13_dav_aktuelles_reichsstaedte.html.

Gebel, T. (2018). *Freie Privatstädte. Mehr Wettbewerb im wichtigsten Markt der Welt*. Walldorf: Aquila Urbis Verlag.

Gelles, D. (2017). Floating cities, no longer science fiction, begin to take shape. *New York Times*, 13 November. Retrieved from www.nytimes.com/2017/11/13/business/dealbook/seasteading-floating-cities.html.

Government of Honduras (2013). Decreto No. 120–2013. Ley Orgánica de las Zonas de Empleo y Desarrollo Económico, (ZEDE), *La Gaceta, Diario Oficial de la Republica de Honduras*, 6 September, Seccion A, No. 33.222, A.

Government of Honduras (2014). Acuerdo Ejecutivo No. 003–2014, *La Gaceta, Diario Oficial de la Republica de Honduras*, 1 January, Seccion A, No. 33.342, A.

Hirschman, A. O. (1970). *Exit, voice, and loyalty. Responses to decline in firms, organizations, and states.* Cambridge, MA: Harvard University Press.

Igarapé Institute (2019). Honduras. *Homicide monitor.* Retrieved from https://homicide.igarape.org.br/.

Instituto Nacional de Estadística, Honduras (2008). *Encuesta agrícola nacional 2007–2008. Tenencia, Uso de la Tierra, Crédito y Asistencia Técnica,* Tegucigalpa.

La Tribuna (2018). '$400 millones aumentará la inversión con las ZEDEs': SDE. *La Tribuna,* 12 October. Retrieved from www.latribuna.hn/2018/10/12/400-millones-aumentara-la-inversion-con-las-ZEDEs-sde/.

Liasson, M. (2001). Conservative Advocate. *National Public Radio – Morning Edition,* 25 May. Retrieved from www.npr.org/templates/story/story.php?storyId=1123439?storyId=1123439&t=1550078584664.

Lynch, C. R. (2017). 'Vote with your feet': Neoliberalism, the democratic nation-state, and utopian enclave libertarianism. *Political Geography, 59,* 82–91.

Österreichische Nationalbank (2019). *Dr. Barbara Kolm. Vice President, Oesterreichische Nationalbank.* Retrieved from www.oenb.at/en/About-Us/Organization/Decision-Making-Bodies/General-Council/vice-president-barbara-kolm.html.

Quirk, J., & Friedman, P. (2017). *Seasteading. How floating nations will restore the environment, enrich the poor, cure the sick, and liberate humanity from politicians.* New York: Free Press.

Robinson, M. (2018). An island nation that told a libertarian 'seasteading' group it could build a floating city has pulled out of the deal. *Business Insider,* 14 March. Retrieved from www.businessinsider.com/french-polynesia-ends-agreement-.

Romer, P. (2009). Why the world needs charter cities. *TED.* Retrieved from www.ted.com/talks/paul_romer?language=en.

Romer, P (2010). Technologies, rules, and progress: The case for charter cities. *Center for Global Development.* Retrieved from www.files.ethz.ch/isn/113646/1423916_file_TechnologyRulesProgress_FINAL.pdf.

Ruchlak, N. (2014). Interview with Michael Strong, 30 May.

Ruchlak, N. (2015). Interview with Barbara Kolm, 4 September.

Seasteading Institute (2019a) Staff/board/advisers. *The Seasteading Institute.* Retrieved from www.seasteading.org/about/staff-board-advisors/.

Seasteading Institute (2019b). Vision/Strategy. *The Seasteading Institute.* Retrieved from www.seasteading.org/about/vision-strategy/.

Seasteading Institute (2019c). The eight great moral imperatives. *The Seasteading Institute.* Retrieved from www.seasteading.org/videos/the-eight-great-moral-imperatives/.

Steinberg, P. E., Nyman, E., & Caraccioli, M. J. (2012). Atlas swam: Freedom, capital, and floating sovereignties in the seasteading vision. *Antipode, 44*(4), 1532–1550.

Strong, M. (2017). How seasteading will enrich the poor. *The Seasteading Institute.* Retrieved from www.seasteading.org/podcast-michael-strong-how-seasteading-will-enrich-the-poor/.

Woolf, N., & Wong, J. C. (2016). Peter Thiel goes 'big league', joining Trump's presidential transition team. *Guardian,* 11 November. Retrieved from www.theguardian.com/technology/2016/nov/11/peter-thiel-joins-donald-trump-transition-team.

World Bank (2018). The World Bank in Honduras. Overview. *The World Bank.* Retrieved from www.worldbank.org/en/country/honduras/overview.

Yarow, J. (2013). Google CEO Larry Page wants a totally separate world where tech companies can conduct experiments on people. *Business Insider,* 16 May. Retrieved from www.businessinsider.com/google-ceo-larry-page-wants-a-place-for-experiments-2013-5?op=1&IR=T.

Part II
Governing cities in neo-liberalism

4 Restructuring Melbourne
Uneven geographies of success

Seamus O'Hanlon

Introduction

In common with the experiences of a number of other former manufacturing cities worldwide, the last quarter of the twentieth century was not kind to Australia's second-largest city, Melbourne (High, MacKinnon, & Perchard, 2017; Dingle & O'Hanlon, 2009; O'Hanlon, 2018). A sprawling suburban metropolitan region with a post-war economy based on low-value-added and tariff-protected manufacturing, the onset of post-Fordist economic restructuring brought a series of crises to Melbourne from the late 1970s through to the mid-1990s. The first of these was centred on the high-profile inner region which experienced rapid deindustrialisation in the late 1970s as older industries associated with manufacturing household consumer goods and textiles, clothing, and footwear (TCF) products, as well as printing, food processing, and the production of light-engineering goods, moved to larger and more modern facilities on the urban fringe or, more commonly, relocated production off-shore to the newly-industrialising countries of Asia. Later, in the 1980s and 1990s, largely as an outcome of globalisation and national policies designed to open up the Australian economy to international competition and to 'enmesh' the country into the growing economies of Asia, industries that had thrived in the post-war Fordist years such as heavy engineering, furniture, and appliance-manufacturing and similar consumer-goods industries, most of which were based in the middle and outer suburban districts of the metropolitan area were similarly impacted (Bongiorno, 2017; O'Hanlon, 2018). Further restructuring in the 2000s and beyond saw the largely Melbourne-based Australian auto industry and its local supply chain downsized before being entirely shuttered in 2016 and 2017.

Together, the effects of these changes have seen manufacturing employment in the Melbourne metropolitan area decline from more than 30 per cent of the workforce in the early 1970s to less than 10 per cent today (Australian Bureau of Statistics, 2016). That figure drops to less than 4 per cent in the increasingly gentrified inner city. Across the metropolis, the largest single industry grouping is now 'health care and social assistance', which accounts for 12 per cent of all employees, followed by retail trade with 10 per cent. In the inner city the largest employment group is 'professional, scientific, and technical services', which account for just under 16 per cent of all employees. More than 40 per cent of all employees in metropolitan Melbourne are now 'professionals or managers', a figure that rises to 50 per cent in the gentrified inner zone. As such, while manufacturing employment has radically contracted, both numerically and relative to other industries in Melbourne since the 1970s, unlike in some other former manufacturing centres in North America and Europe, which have never recovered from the effects of deindustrialisation and economic restructuring, in recent years Melbourne's economy has rebounded and the city is now enjoying unprecedented prosperity. One manifestation of that prosperity is rapid population growth, which has seen the metropolitan region grow by more than 100,000 residents per annum for the last decade and by more than 1.5 million people since the turn of the century. Across the four decades since the mid-1970s the metropolitan population has doubled from just over 2.5 million people to more than five million (Australian Bureau of Statistics, 2018). While much of this growth has been on the urban fringe, significant redevelopment and repopulation has also occurred at the centre, in neighbourhoods that in the 1970s were characterised by high unemployment and rapid population decline. In the post-Fordist era, then, Melbourne's metropolitan region has both deindustrialised and doubled in size.

It has also become extraordinarily ethnically-diverse, expanding on and extending a process that began in the Fordist years when the metropolitan area also doubled in size partly because of the well-documented post-war baby boom, but mainly as a result of elevated levels of immigration, mostly from Europe. These immigrants were enticed to the city to work in the rapidly expanding manufacturing sector, in many cases sponsored by employers and governments' intent on growing the population for economic and security reasons. As a result of this immigration within a generation after World War II Melbourne evolved from being an essentially native-born and mono-cultural city, dominated by the ethnically British and Irish, to an immigrant one. By the early 1970s more than 40 per cent of Melbourne's population were immigrants or their children, up from less than 10 per cent after the war. The city also became multicultural as significant numbers of these new arrivals were from Southern and Eastern Europe. Others were from diasporic European communities in the Middle East, Africa, and Asia (O'Hanlon & Stevens, 2017). And while by the 1970s just over 30 per cent of Melbourne's workforce were employed in manufacturing, neither these workers nor their jobs were evenly spread across the metropolis. Instead, they tended to congregate near their jobs in the city's manufacturing zones such as the traditional home of the shoe-making industry in inner-urban Collingwood where nearly half of the workforce were both immigrants and manufacturing workers (Dingle & O'Hanlon, 2009). In outer suburban Dandenong which, as we shall see, was both a major immigrant destination and a significant manufacturing region at that period, 39 per cent of the population were manufacturing workers.

In the decades since, while manufacturing employment has declined rapidly in Melbourne, immigrant settlement has not. In that time the city has, as in the post-war period, again witnessed profound demographic change and has in the process become poly-ethnic. Changes in national immigration policy in 1973 which removed the last vestiges of racially discriminatory elements stemming from colonial times and the early years of Federation mean that immigrants now come from a vast array of source countries and ethnicities, rather than as before, mainly from Europe and European diaspora communities. At the time of the 2016 Census just over 40 per cent of the Melbourne metropolitan population (more than 1.8 million people) was foreign-born, and almost a quarter more had at least one overseas-born parent. Demographers Marie Price and Lisa Benton-Short of the Global Urban Migration (GUM) project at George Washington University have recognised Melbourne as an important global 'immigrant gateway' city, one of only 22 metropolitan regions worldwide with more than one million foreign-born residents. They also classify it as one of the world's most 'hyperdiverse' cities because of the extreme range of residents' birthplace (Price & Benton-Short, 2007).

This 'hyper-diversity' is borne out by a recent survey by Multicultural Affairs Victoria which found that at the time of the 2016 Census residents of the state of Victoria (of which Melbourne is the capital) claimed 247 separate birthplaces, a figure that does not include more than 5,000 overseas-born whose birthplaces were 'inadequately described', nor the more than 400,000 whose country of birth was 'not stated'. Nor does it adequately capture the ethnicity of those who for whatever reason do not identify their national, ethnic, or religious identity with their place of birth (Lobo, 2010). The same report noted that 234 languages (broadly defined) were spoken in the state, while 135 religions were practised (Multicultural Affairs Victoria, 2018). Most of these foreign-born residents lived in metropolitan Melbourne rather than rural and regional Victoria, and increasingly in the outer suburbs of the metropolis rather than the inner city, as was the case until the 1970s. As with many 'immigrant gateway' cities internationally metropolitan Melbourne is now increasingly demographically and ethnically distinct from both its state hinterland, and more so from the rest of Australia. Melbourne (along with Sydney) is increasingly foreign-born and ethnically and religiously diverse, while non-metropolitan Australia is largely local-born and of Anglo and Celtic origin (Colebatch, 2017; O'Hanlon & Stevens, 2017).

These two trends – deindustrialisation and mass immigration – suggest that metropolitan Melbourne can arguably be said to have successfully transitioned from an industrial to post-industrial society over the last four decades, and indeed this has become the dominant narrative of the city from local and state governments, business lobbyists, and others (Gehl People, n.d.; Committee for Melbourne, n.d.; Office of the Victorian Premier, 2019). The city can also plausibly lay claim to having successfully emerged as an exemplar of multiculturalism, as arguably one of the world's great 'hyperdiverse' city-regions. But this macro story of success masks many contemporary problems and glosses over a series of crises that scarred the city in the 1970s, 1980s, and 1990s. The rest of this chapter, then, is a historically informed study of Melbourne's fortunes from the 1970s through to today that shows that the processes of economic and demographic change in Melbourne, as elsewhere, has been neither a simple nor a straightforward story of success and succession. Cognisant of Baker's (2003) dictum that studies of the urban must include a temporal alongside a spatial focus, this chapter seeks to do two things. First, I chronical how, in response to a series of economic crises from the late 1970s through to the 1990s, civic and business leaders in Melbourne drew on narratives about globalisation and global competitiveness to pursue an agenda designed to recreate the city in conformance with neo-liberal, free-market ideas about post-industrial urban activities and forms. Second, I demonstrate that the outcomes of these ideas and policies have not been evenly shared across the metropolis.

In order to do this I document the stories of multiple past and present Melbournes, and present two case studies of change to show how one Melbourne, the increasingly prosperous and gentrified post-industrial inner region – the area that encompasses an approximately 10-kilometre arc circling the Central Business District (CBD) – has benefitted greatly from many of the changes that have come with neo-liberalism and the growth of a service-based economy. These include increased property wealth, better access to well-paid employment opportunities, improved public and private transport services, as well as upgraded cultural and sporting facilities, and what are increasingly called the 'lifestyle opportunities' of inner-city living.

The second case study then looks at another area whose experience of restructuring has been less sanguine, the municipality of Greater Dandenong and its surrounds in the outer southeast of the metropolitan area. The former home of auto plants and other large-scale manufacturing industries, Greater Dandenong has not fared so well in the post-industrial era, although as we shall see it has emerged in recent decades as one of the most ethnically-diverse urban regions in the world. In order to achieve these aims the chapter draws on historical and contemporary census material, a series of historical and contemporary local and municipal reports on the two regions, a classic 1970s era sociological study, and a more recent follow-up one of the outer suburban region, as well as recent state and local government planning and economic development proposals for both areas.

Restructuring and reinventing inner Melbourne

The 1970s began with a note of optimism about the future in Melbourne, including its inner region. Unemployment was low, inward and domestic investment was strong, and the city's port was, as it had been for over 100 years, the main entry point for visitors, goods, and migrants to Australia. New cultural ideas emerged from inner Melbourne-based musicians, novelists, and playwrights in the early 1970s, while movies and television series set in and around the city centre heralded a revival of the local and national film industry (Garner, 2004; Wolf, 2008). But while the city remained the business, manufacturing, and cultural capital of Australia, behind this façade of prosperity lay some major structural problems that were to come to the surface in the middle years of the decade. The most important of these was the over-concentration in Melbourne, and especially its inner region, of the nation's tariff-protected manufacturing industries, which in a report published in 1975 were described as being mired in a 'deep-seated and long-standing malaise', characterised by a low investment, poor conditions, and inward-looking and uncompetitive work and management practices (Committee to Advice on Policies for Manufacturing Industries, 1975, p. 1).

Australia's other major city, Sydney, was also home to a disproportionately large share of the nation's manufacturing sector, but by the late 1970s it began to emerge as the nation's premier economic, tourism, and immigrant 'gateway'. Long home to the country's largest airport, the growing prevalence of air rather than ship travel from the mid-1970s onwards saw Sydney become by far the most important entry point to the country for business visitors, tourists, and immigrants. This phenomenon was to become increasingly prevalent as the nation embraced globalisation in the 1980s and beyond. Sydney's emergence as Australia's leading city was perhaps best exemplified by the opening of its Opera House in 1973 which almost immediately became one of the key international symbols of the country (Spearritt, 1999). Melbourne's loss of economic and cultural importance was also exemplified by the decisions of many of the new and emerging 'sunrise industries' of the globalising era – banking and finance, technology, media and communications, tourism, and property development – to locate their headquarters in Sydney. A number of major Melbourne-based companies were also taken over by Sydney-based ones in the 1980s, or were absorbed by global corporations, many of whom also based their Australian (or Asia-Pacific) corporate headquarters there. As such in the 1980s and 1990s Melbourne increasingly came to resemble a 'branch-office' city. It also began to decline economically relative not only to Sydney but also to the emerging cities of Brisbane and Perth. An Australian version of what is called 'sunbelt' migration in the US saw Perth grow rapidly on the back of a mining boom in the state's northwest in the 1970s and 1980s, while in Queensland to the north mass tourism and retirement migration saw a similar population drift to sub-tropical Brisbane and nearby coastal regions.

The onset of the international recession in 1974–1975 and the return of unemployment rates not seen since the 1930s exposed many of Melbourne's problems, especially its over-reliance on the highly-protected TCF sector which was largely based in inner Melbourne. As in inner cities elsewhere in the world at this time not only were jobs being lost in inner Melbourne, so too was population. This decline both nominal and in proportion to the rest of the metropolitan area, a problem that was to continue well into the 1980s. While in 1971 the population of Melbourne's 'core', defined by the Melbourne and Metropolitan Board of Works (MMBW, the nearest equivalent Melbourne had to a metropolitan municipal government) as the municipal City of Melbourne and its seven immediate neighbours was just under 310,000 – which was about 12 per cent of the total metropolitan population. Five years later that number had dropped by more than 15 per cent (or 50,000 people) to just under 260,000 (MMBW, 1977a). By 1981 it was just 245,000. Melbourne's inner city thus lost more than 20 per cent of its population in the single decade of the 1970s, with most of these leaving in the five years to 1976. By the time of the 1986 Census another 14,000 had left, leaving a population of about 231,000, before stabilising at around that number in the early 1990s (Dingle & O'Hanlon, 2009). Perhaps more tellingly by the mid-1980s the inner core had become a tiny fraction of the metropolitan population, home to less than 10 per cent of the city's then more than three million people.

As population and jobs continued to drift to the suburbs, to Sydney, to the 'sunbelt', and overseas in the 1970s and 1980s concerns about the ongoing economic viability of inner Melbourne became a real topic of concern in business and government circles. These concerns in turn became a catalyst for many of the changes that have been instituted in the decades since. The closures of factories were abrupt and highly-visible examples of the effects of the economic restructuring of the inner city, but so too the rapid closure of a number of major city centre and inner-urban department stores in the wake of the 1974–1975 recession acted as a harbinger of major economic and social change. As population and employment suburbanised so too did shopping and leisure, exemplified by the growth of drive-in suburban shopping centres such as the city's first Chadstone, which opened in 1960. It was followed in the 1970s and 1980s by other major centres to the north, south, east, and west of the city centre. The collapse of a commercial property boom in the CBD and then a move away from what was called 'comprehensive redevelopment' in the face of resident protest and changing architectural and planning fashions in the mid-1970s also left the inner city pockmarked with numerous semi-derelict blocks of empty housing awaiting its fate. Similarly, the abandonment of the city's docks and warehouse districts directly adjacent to the downtown area, in response to technological change and the containerisation of shipping, saw the inner city develop a rundown and abandoned feel. And while at that stage gentrification was beginning to have an effect on the demography of the inner city, for the most part until the late-1970s inner Melbourne remained a place of working-class and immigrant employment and residence (Logan, 1985; Howe, Nichols & Davison, 2014). It was these communities who bore the brunt of economic restructuring and who witnessed the physical deterioration of their neighbourhoods in the 1970s and 1980s before rejuvenation occurred in the 1980s and beyond.

Combined, these developments created a strong sense for many Melburnians that their city was facing decline, with its best days possibly behind it. This concern was both popularly felt and became pervasive in government and corporate circles. Two reports published by the planning arm of the MMBW in 1977 confirmed this sense of malaise and crisis, especially in inner Melbourne. While demonstrating that the inner region remained by far the most important employment zone in the metropolitan area, census and other figures compiled by the MMBW showed that it was rapidly losing jobs and population to the outer suburbs and off-shore (MMBW, 1977a). After detailing some of the problems facing the inner city as manufacturing jobs began to decline in the face of international competition, automation, and technological change – as well as associated problems such as high rates of unemployment and poverty, degraded urban environments amongst others – these reports suggested that if nothing was done to arrest inner Melbourne's economic and social problems there existed the real possibility that the region might face a bleak hollowed-out future of mass unemployment and urban rioting similar to that increasingly being seen in the US and the UK (MMBW, 1977b).

Social democratic politician Rob Jolly (2008) who became state treasurer in the midst of the 1982 recession recalls that he and his colleagues believed that Melbourne seemed to be 'moribund' trapped in what seemed to be a cycle of economic and social decline, dangerously reliant on tariff-protected but slowly dying manufacturing industries. As with the MMBW, he saw the economic and physical deterioration of the inner city as both symptomatic of and exacerbating a sense of metropolitan decline. His immediate response and that of most of his successors, as well as many of the city's business and social elites in the decades since, has been to attempt to reinvent Melbourne as a post-industrial city with a leisure- and services-oriented future rather than a manufacturing one with a fading past. As in the 1980s his government worked with business and other groups to support new 'sunrise industries', such as finance, funds management, technology and logistics, tourism, and the construction industry, often using public funds to support joint ventures in these fields.

More overtly Jolly's government (and its successors) have invested huge amounts of public money in urban infrastructure projects, most notably the rejuvenation of the city's major sporting and cultural assets, almost all of which are located within the increasingly gentrified inner city (O'Hanlon, 2009; Government of Victoria, 1984). The aim was to turn the inner city into a place where leisure, business services, and education would become major post-industrial competitive strengths. A more overtly neo-liberal government elected in the wake of the early 1990s recession when unemployment peaked at over 13 per cent in metropolitan Melbourne, saw the rejuvenation of the inner city's cultural infrastructure as an essential component of a plan to enhance 'Melbourne's attractiveness as a lively cosmopolitan centre with a vibrant retail, entertainment and cultural focus'. As such the new government's 'Agenda 21' initiative sought to 'strengthen the City's key financial and commercial activities' by expanding the capital city as a base for trade and advanced manufacturing and research, but more importantly by 'promoting Melbourne as a showcase of world-class events and festivals, as the home of the arts, sports and conventions within Australia and within the Asia-Pacific region' (Victoria. Office of Major Projects, 1995, p. 2).

This now more than three-decade-long 'events' strategy of re-imagining and rebuilding inner Melbourne along post-industrial lines as a way of achieving success in the neo-liberal globalising era has been a great success. Similar to a number of other former manufacturing cities across the globe such as Pittsburgh, today's inner Melbourne bears little resemblance to its rather grim 1970s and 1980s self (Neumann, 2016). Across the inner city, formerly industrially and economically redundant spaces such as the south bank of the Yarra River, the former docks area immediately west of the CBD and former factory areas to its north and north-west have been rebuilt as commercial, residential, and leisure zones featuring convention centres, hotels, shopping centres, and hospitality precincts. The broader inner region has also reversed the de-population of the 1970s and 1980s and is now experiencing rapid population growth, manifested in numerous high-rise residential developments and urban renewal projects.

Nearly one million people now visit the centre of Melbourne every day, with that figure expected to grow by 40 per cent in the next 20 years, while the population of the roughly 40 square kilometre area of the municipal City of Melbourne has more than doubled in the last 25 years. That growth has been in spite of efforts of the state government to use a municipal restructure in the mid-1990s to privilege business interests in the operations of the council by excising traditional inner-urban residential neighbourhoods from the council's boundaries. Approximately 45,000 people now live in the CBD (downtown area) whereas only a few hundred did so in the early 1990s. A residential apartment building boom is expected to see that population grow by another 20,000 by 2020. As such, Melbourne's CBD is now one of the fastest-growing residential regions in Australia (City of Melbourne, 2018). The broader inner Melbourne region has gained well over 100,000 new residents over the same period, with tens of thousands of these international students, mostly from China and India, who simply did not exist as a feature of Melbourne life until the late 1980s when the provision of international education became one of the new 'sunrise industries' championed by the state and national governments. In recent decades federal, state, and local governments as well as universities and businesses have invested huge amounts in education and research facilities drawing on the region's long tradition as a place of excellence. From almost no international enrolments in the mid-1980s, Melbourne is now recognised as one of the most popular international student destinations in the world, home to more than 200,000 at any one time. Most of these students live in the city's inner region, with the municipal City of Melbourne alone home to approximately 50,000 (City of Melbourne, 2018; O'Hanlon, 2018).

The inner city also now attracts a growing number of tourists, both domestic and international, who come to enjoy a now almost weekly calendar of major sporting, cultural, and community 'events' most of which have been invented since the mid-1980s, largely by local authorities and local business groups seeking to bring back customers to what were sometimes dying retail strips by 'branding' them ethnically or culturally (O'Hanlon, 2009, 2018). While Melbourne is usually rated as 'beta' or 'gamma' in various 'global city' rankings, since 2010 it was been either the winner or runner up in the *Economist* magazine's Intelligence Unit's annual 'most liveable' index which measures the qualities that most appeal to expatriate business people and their families, who for the most part live, work, and socialise in the now gentrified inner city

(Economist/Economist Intelligence Unit, 2018). Jan Gehl, the Danish architect and planner who has had a role in these processes of the re-invention of Melbourne through a series of planning and 'place-making' interventions into the urban fabric, has described the city's re-invention as a 'miracle' (Gehl People, n.d.). In many ways he is right, inner Melbourne has been rejuvenated, but as with many reinvented former manufacturing cities internationally it is important to recognise that the new prosperity of the post-industrial neo-liberal era is unevenly shared across the metropolis.

Restructuring and reinventing Greater Dandenong

While the inner city has emerged as an important and inviting post-industrial region in recent years, in other parts of the metropolitan area restructuring has been a far more painful process that has not (yet?) had a happy ending. As in other restructured cities across the globe, Melbourne's post-industrial success story has been distinctly uneven, with a number of former manufacturing areas missing out on many of the benefits of the new prosperity. One area where this has especially been the case is the City of Greater Dandenong, which lies about 30 kilometres south-east of central Melbourne. Created as a political entity in a restructuring of municipal boundaries in 1994, Greater Dandenong is based around a former market town which expanded rapidly in the 1950s and 1960s when it was absorbed into the Melbourne metropolitan area and became a major a site of manufacturing. General Motors, in the form of its Australian affiliate General Motors Holden (GMH), established a major manufacturing plant there in the mid-1950s, as did International Harvester and HJ Heinz, amongst others. These plants were located just outside the old market town on a major highway which linked it to the wider metropolitan area. It was also serviced by a regional freight and commuter railway line, including a new stop on the suburban railway network (named General Motors) which was created to provide employee access to the site in 1956. These facilities, along with guarantees of tariff protection, were provided by state and federal governments as incentives for the companies to establish local operations in Australia. The Victorian state government also provided subsidised housing for workers from the factories and their families in a nearby public housing estate called Doveton, built by its public housing authority the Housing Commission of Victoria (HCV) in stages beginning in 1956.

There were strong echoes of Ford's investment in interwar Dagenham in outer London in the Doveton story, and indeed as was the case there Doveton was the subject of a major urban sociological study carried out in the late 1960s. Explicitly modelled on Young and Wilmot's (1957) famous *Family and Kinship in East London*, which examined community and community formation in Bethnal Green in inner London and the new suburban community of Dagenham in the early post-war years, *An Australian Newtown: Life and leadership in a working-class suburb* by Lois Bryson and Faith Thompson has become something of a classic in Australian urban studies (Young & Wilmott, 1957;

Bryson & Thompson, 1972). Its key findings were that while Doveton lacked many services such as community facilities, hospitals, and parks for recreation, most of its residents were content with their lives. This was largely because it was a place of full (male) employment and because many of the residents were recent arrivals to the outer south-eastern metropolitan region – either local-born Melburnians from rundown inner-city neighbourhoods, or more commonly immigrants from poorer parts of Britain, Ireland, and continental Europe. Compared to their former addresses their homes in Doveton offered space, sunshine, and privacy.

Like Dandenong more generally, Doveton was a strongly working-class community. Most of the men were employed in manufacturing while the women either stayed at home and looked after children or similarly worked in the local factories, albeit in different jobs than the men. *An Australian Newtown's* researchers found that fully 75 per cent of Doveton's men were employed in unskilled, semi-skilled, and skilled manual work while less than 5 per cent were professionals; the researchers could find no doctors, lawyers, lecturers, or business owners, and only five high school teachers amongst their sample population of 324 male heads of households. Of the 339 married women in their sample 103 remained in paid employment after marriage, but only five worked in professional jobs. Like Dagenham, Doveton was a classic working-class Fordist community (Bryson & Thompson, 1972). One of its residents was Dennis Glover, whose family were immigrants from Ireland and England who moved to the suburb in the late 1960s (Glover, 2015). Both Glover's parents worked at GMH, his father on the production line and his mother in the canteen. Like a number of his neighbours Glover also worked there as a young man, in his case as a part-timer during his university holidays; as with many millions of others across the West during the Fordist 1960s and 1970s he was able to take advantage of the 'post-war settlement' to get an education and move into a different class than his parents. He ended up with a PhD in History and various jobs in the private sector, government, and politics.

One of the authors of the original *Newtown* study, sociologist Lois Bryson returned to Doveton in the early 1990s to report on 'social change and suburban lives' from the 1960s to the 1990s. Interviews conducted with some original participants and many new ones in the wake of post-Fordist restructuring processes and in the midst of the early 1990s recession to report made for grim reading (Bryson & Winter, 1999). Even more confronting was a memoir published in 2015 by Dennis Glover about growing up in Doveton and what has happened to it and other places like it since the end of the long post-war boom. His verdict on the contemporary situation is scathing:

> The entrance to the old GMH factory in Dandenong is now a public road, 'Assembly Drive'; half-finished and unsealed in places, it looks like the last street in a county town. The vast flat stretches of tarmac, which thousands of gleaming new vehicles once covered, are now empty, weeds growing through cracks (…). They are now in the process of being replaced by

warehouses of bolted concrete and steel sheeting (…) the old factory buildings have disappeared. This, you imagine, is how archaeologists reconstruct the sites of former Greek settlements along the Mediterranean – the long shadows at morning and sunset revealing the sites of razed markets and temples, the retreat of the sea showing a once thriving Phoenician port.

(Glover, 2015, p. 14)

As is obvious from this excerpt, while like the rest of Melbourne the Greater Dandenong regional economy has been restructured in the post-Fordist era its contemporary story bears little resemblance to that of the success story of inner Melbourne.

Also, unlike in inner Melbourne, in Greater Dandenong, today manufacturing remains an important component of the local economy. Indeed it is the largest single employment group, but unlike in the inner city and in the full-employment Fordist period, a sizeable portion of Greater Dandenong's residents have essentially been declared surplus to the needs of the post-industrial economy. Alongside manufacturing, large numbers of locals are employed in wholesaling and retailing, health and social care, and construction, but in both Dandenong itself and in former working-class regions in its vicinity, for many, and especially older men, unemployment benefits, the disability pension, or early retirement are now amongst their major fields of 'employment'. Today's Doveton's residents remain, as in Dennis Glover's day, low-income earners, but rather than low-paid working people like Dennis' family and his former neighbours they are increasingly low-paid non-working people. The factories have gone from Greater Dandenong and Doveton, along with many of the unskilled jobs they once offered, but unlike in inner Melbourne these have not been replaced with post-industrial ones. Of the 4,000 males aged 15 years and over who lived in Doveton at the time of the 2016 Census over 13 per cent were unemployed (double the metropolitan Melbourne rate of 6.5 per cent), but even this elevated figure hides the full extent of the area's problems; only 50 per cent of men in the prime of their working lives (aged 44–64) were in the workforce and thus counted amongst the unemployed and actively looking for work. Large numbers of Doveton's men are now either outside the workforce having taken voluntary or involuntary early retirement, or more commonly on disability and other pensions.

For women the unemployment rate was slightly higher at 14 per cent, but in contrast to their menfolk more than 60 per cent of the female working-age cohort was actually in the workforce. The largest single employment category for these women was 'health care and social assistance', which was also the largest employment sector for all women in Doveton, responsible for nearly twice the number of jobs than the second-largest sector, retail trade (Australian Bureau of Statistics, 2016). As these figures suggest, in Doveton, as in many other post-industrial places internationally, economic restructuring has been very tough on the whole community, but it is often older unskilled men who have borne the brunt of job losses and whose anger is so often reflected in the literature on deindustrialisation (High, 2013; High, MacKinnon, & Perchard, 2017).

Like inner Melbourne the wider Greater Dandenong region is today a very different place from what it was 50 years ago at the end of the Fordist era. And while Doveton and places like it have major economic and social problems that must be addressed if everyone is to share the prosperity of the post-industrial era, it is misleading to characterise these places as economically unproductive or socially and culturally moribund. Greater Dandenong may have lost its manufacturing base; however, in the last 40 years, it has developed an extraordinary entrepreneurial culture in the form of a massive and rapidly growing immigrant population. Greater Dandenong generally, and specific areas within it are now amongst the most multicultural urban regions in the world (Tomaney, 2015; Lobo, 2010). Across the municipality almost three-quarters of residents are foreign-born and only 12 per cent have two Australian-born parents (Australian Bureau of Statistics, 2016). Within the region there are neighbourhoods that have been demographically, economically, and culturally transformed in recent decades by settlers from Vietnam, China, India, and Afghanistan and more recently from the Horn of Africa and sub-Saharan Africa. As in Melbourne more generally there are nearly 250 different ethnic groups represented in Greater Dandenong, communities who have exhibited an extraordinary entrepreneurial spirit and who represent a great local competitive strength.

Municipal authorities in Greater Dandenong increasingly recognise this diversity and see multiculturalism and immigrant business districts as important economic resources in the post-industrial era. As such the local government website now actively promotes the openness of the municipality to newcomers (especially refugees and asylum seekers) and the impacts these communities have had and continue to have on particular shopping streets and districts. As such these people and places are presented as key economic and social strengths of the 'new' Dandenong. One local initiative, dubbed 'Hello' highlights the cultural heritage of neighbourhoods such as 'Sensational Springvale', 'melting pot of Asian cultures (...) that could easily have you thinking you are in South-east Asia rather than suburban Melbourne'. It also highlights cultural tourism as an element of the contemporary Dandenong economy. 'Springvale Fresh' a biennial food tour, which alongside the 'Dumpling and Desserts' tour of the same neighbourhood, are key components of a municipal-sponsored initiative to make Greater Dandenong a 'must-go destination for people who love food and culture'. Similarly, 'Little India' in Central Dandenong's Foster Street, which was recently officially recognised as 'Victoria's first-ever Indian cultural precinct', and is actively marketed by the municipality as 'Melbourne's longest-standing and most authentic cluster of Indian and sub-continental culture and commerce'. Another stop on a 'food lovers tour of Dandenong' the council's website informs us that 'there are approximately 30 shops in Little India featuring speciality goods from India, Pakistan, Fiji, Sri Lanka, and Bangladesh, servicing an Indian catchment of some 85,000 people'. Nearby Thomas Street is also now known as the 'Afghan Bazaar', 'one of Melbourne's most prominent cultural precincts that highlights the intense concentrations of Afghan traders' (City of Greater Dandenong, 2018a, 2018b, 2018c).

But as with other governments at local, state, and national level, and in a reflection of many of the attitudes that remain over from the Fordist era, in Greater Dandenong there remains a tendency to look to big projects and to big external investors rather than local initiatives as drivers of future prosperity. Whereas municipal, state, and national governments once sought out foreign investment in manufacturing to drive economic development, today the emphasis is on major construction projects and the services sector. An example of this is a recent initiative from the local council called the 'Revitalising Central Dandenong Project', which is seeking to rebuild the former site of the market, redundant factories and other major structures knocked down during a mania for comprehensive redevelopment in the 1980s and 1990s with large-scale civic facilities, multi-storey office buildings, and high-rise apartment blocks, as well as hotels, hospitality and convention facilities (City of Greater Dandenong, 2018b). While some of these are the things that worked in revitalising inner Melbourne in the post-Fordist era they likely to have little or no chance of such success in Dandenong, largely because the locals there do not have the money to patronise them and the area attracts few high-paying tourists other than business people visiting nearby business and office parks. Nor does it have the long-standing tertiary education and research facilities that have been so central to inner Melbourne's post-industrial resurgence.

Conclusion

Greater Dandenong, Doveton, and places like them, do have opportunities to create successful futures, but these will very likely not be found in major building projects or trophy investments, but from efforts by local, state, and national governments to again invest in their own people to enable them to better themselves and to provide a better future for their children and grandchildren – just as they did in Dennis Glover's time. There may be a case for doing this through major planning initiatives like 'Revitalising Central Dandenong', or by seeking to attract footloose multinational capital investment, but more likely it will come from smaller-scale changes in which local people are encouraged and supported to invest in small businesses, when diversity and cultural endeavour is celebrated and when people have a sense that Dandenong is a place for everyone not just those who can afford to buy private facilities and services. Future success in places like Dandenong will more likely come from business and cultural initiatives such as promoting businesses in Springvale, the Afghan Bazaar, and Little India than from big initiatives.

While the revitalisation of post-Fordist inner Melbourne owed much to major building projects and to large-scale government initiatives and spending, it also owed a lot to the idea that cultivating and celebrating the talent and the increasing diversity of the city's population and a policy of allowing residents – old and new – to be themselves, maintain their cultural distinctiveness, and to follow their own paths, was a major competitive advantage in the post-industrial era. One of the keys to urban success in the contemporary world is to celebrate difference and to trust people and local communities to find their own futures.

And while this might sound neo-liberal, it does not have to be. In order to allow people and communities to succeed we must ensure that they have the financial and other capacities to do so. Market solutions have the capacity to allow this, but success is more likely to come from entrepreneurialism backed by a social safety net, just as it was in the Fordist era. Into the future successful cities and regions are likely to be those that remain open to new people and new ideas from wherever they originate, allow residents to cultivate and develop their existing links with the world, and find local solutions to local problems rather than import off-the-shelf ideas from somewhere else. In the post-industrial era successful cities and regions will be those that are themselves and comfortably so, not those that try to be what they are not and never can be.

References

Australian Bureau of Statistics (ABS). (2016). *Census of population and housing*. Canberra: Australian Bureau of Statistics.

Australian Bureau of Statistics (ABS). (2018). *Regional population growth, Australia, 2016–17*. (3218.0). Canberra: Australian Bureau of Statistics.

Baker, A. R. H. (2003). *Geography and history: Bridging the divide*. Cambridge: Cambridge University Press.

Bongiorno. F. (2017). *The eighties: The decade that transformed Australia*. Melbourne: Black Inc.

Bryson, L., & Thompson, F. (1972). *An Australian Newtown: Life and leadership in a new housing estate*. Melbourne: Penguin.

Bryson, L., & Winter, I. (1999). *Social change, suburban lives: An Australian Newtown, 1960s to 1900s*. Sydney: Allen and Unwin.

City of Greater Dandenong (2018a). *Neighbourhood shopping precincts*. Retrieved from www.greaterdandenong.com/document/18968/neighbourhood-shopping-precincts.

City of Greater Dandenong (2018b). *Revitalising Central Dandenong*. Retrieved under www.greaterdandenong.com/document/3396/revitalising-central-dandenong-.

City of Greater Dandenong (2018c). *Your community. Diversity*. Retrieved under www.greaterdandenong.com/section/2512/diversity.

City of Melbourne (2018). *Daily population and estimates and forecasts*. Retrieved from www.melbourne.vic.gov.au/about-melbourne/research-and-statistics/city-population/Pages/daily-population-estimates-and-forecasts.aspx

Colebatch, T. (2017). One census: Three stories, *Inside Story*, 5 June. Retrieved from https://insidestory.org.au/one-census-three-stories/.

Committee to Advice on Policies for Manufacturing Industry (Jackson Committee) (1975). *Policies for development of Australian manufacturing*. Canberra: Australian Government Printing Service.

Committee for Melbourne (n.d.). *About us*. Retrieved from http://melbourne.org.au/about-us.

Dingle, T., & O'Hanlon, S. (2009). From manufacturing zone to lifestyle precinct: economic restructuring and social change in inner Melbourne, 1971–2001. *Australian Economic History Review*, 49(1), 52–69.

Economist/Economist Intelligence Unit (2018). *Global Liveability Index. The Economist*. Retrieved from https://pages.eiu.com/rs/753-RIQ-438/images/The_Global_Liveability_Index_2018.pdf.

Garner, B. (2004). Attracting an Audience: Theatre. In P. Yule (Ed.), *Carlton: A history* (pp. 175–188). Melbourne: Melbourne University Press.

Gehl People. (n.d.). *The Melbourne Miracle*. Retrieved from https://gehlpeople.com/cases/melbourne-australia/.

Glover, D. (2015). *An economy is not a society: Winners and losers in the new Australia*. Melbourne: Redback.

Government of Victoria (1984). *Victoria. The Next Step: Economic initiatives and ideas for the 1980s*. Melbourne: Government Printing Service.

High, S. (2013). 'The Wounds of Class': A historiographical reflection on the study of deindustrialization, 1973–2013. *History Compass*, *11*(11), 994–1007.

High, S., MacKinnon, L., & Perchard, A. (Eds.) (2017). *The deindustrialized world: Confronting ruination in postindustrial places*. Toronto: University of British Columbia Press.

Howe, R., Nichols, D., & Davison, G. (2014). *Trendyville: The battle for Australia's inner cities*. Melbourne: Monash University Publishing.

Jolly, R. (2008) Interview with Seamus O'Hanlon and Tony Dingle, Monash University, Melbourne, 28 April.

Lobo, M. (2010). Inter-ethnic understanding and belonging in suburban Melbourne. *Urban Policy and Research*, *28*(1), 85–99.

Logan, W. (1985). *The gentrification of inner Melbourne: A political geography of inner city housing*. Brisbane: University of Queensland Press.

Melbourne and Metropolitan Board of Works (MMBW) (1977a). *Melbourne's inner area: A position statement*. Melbourne: MMBW.

Melbourne and Metropolitan Board of Works (MMBW) (Urban Economic Consultants Pty Ltd) (1977b). *Socio-economic implications of urban development*. Melbourne: MMBW.

Multicultural Affairs Victoria (MAV) (2018). *Population diversity in Victoria: 2016 Census. Local government areas*. Melbourne: Department of Premier and Cabinet.

Neumann, T. (2016). *Remaking the rustbelt: The postindustrial transformation of North America*. Philadelphia, PA: University of Pennsylvania Press.

Office of the Victorian Premier (2019). Victoria's boom economy leads the way, Media Release, 29 January 2019. Retrieved from www.premier.vic.gov.au/victorias-booming-economy-leads-the-way/

O'Hanlon, S. (2009) The events city: Sport, culture and the transformation on Melbourne, 1977–2006. *Urban History Review/Revue d'Histoire Urbaine*, *37*(2), 30–39.

O'Hanlon, S. (2018). *City life: The new urban Australia*. Sydney: New South Publishing.

O'Hanlon, S., & Stevens, R. (2017). A nation of immigrants or a nation of immigrant cities? The urban context of Australian multiculturalism, 1947–2011. *Australian Journal of Politics and History*, *63*(4), 556–571.

Price, M., & Benton-Short, L. (2007). Counting immigrants in cities across the globe. *Migration Policy Institute Online Journal*. Washington, DC: Migration Policy Institute.

Spearritt, P. (1999). *Sydney's century: a history*. Sydney: UNSW Press.

Tomaney, J. (2015). Region and place II: Belonging, *Progress in Human Geography*, *39*(4), 507–516.

Victoria. Office of Major Projects (1995). *Agenda 21 Quarterly*, 8.

Wolf, G. (2008). *Make it Australian: The Australian performing group, the pram factory and new wave theatre*. Sydney: Currency Press.

Young, M., & Wilmott, P. (1957). *Family and kinship in East London*. Harmondsworth: Penguin.

5 Governing through participation

Activation of civil commitment in Berlin's neighbourhoods

Magdalena Otto

Introduction

The activation of commitment in neighbourhoods is established as one approach to solving certain problems on the urban level, such as squalidness in public spaces, lack of care supplies due to limited budgets, or exclusion of minorities. Hence, different NGOs, public-private partnerships, and Corporate Social Responsibility initiatives in urban fields share the narrative of participation, joint commitment, and responsibility in order to buffer discrimination, compensate low budgets, and enhance quality of life through actively shaping the neighbourhoods they are situated in. The invocation of public commitment not only as an opportunity to take part as active members of society but as an act of exercising citizens' necessary responsibility reveals a fundamental transformation of how citizenship is characterised at the present time. Therefore, this chapter raises the question of whether we need to state a shift from right to duty in terms of social inclusion.

Activating citizens is part of the current policy agenda in neo-liberal cities. Usually, neo-liberal restructuring is associated with the entrepreneurial turn in urban governments, privatisation, austerity politics, and so forth. Putting it in a nutshell, Brenner and Theodore (2002, p. 351) state: 'Throughout the advanced capitalist world (…) cities have become strategically crucial geographical arenas in which a variety of neoliberal initiatives (…) have been articulated'. Indeed, even integrating civil society actors into local systems of governance appears as another element in neo-liberal urban development processes. Previous fields of responsibility of the welfare state seem to have become less important, while the activation of civic action is gaining more attention as part of the rescaling process in the neo-liberal city. Neo-liberal globalisation and reorganisation of the nation-state has led to a decline in state-run capacities for redistribution and compensation of social inequalities (Heeg & Rosol, 2007, p. 492). Following this argument, decision-making authority is increasingly shifted to subnational and supranational levels (Brenner, 2004), thereby restructuring urban governance and thus changing the way of policy-making (Hess & Lebuhn, 2014, p. 23). The current form of regulation and politics is characterised as a transformation from government to governance (e.g. Evans, Richmond, & Shields, 2005, p. 77; Swyngedouw, 2005, p. 1993).

This new mode of regulation can be observed in the triangle of state, economy, and civil society, especially in the heightened overlap between these spheres. Assuming civil society as a third societal sphere besides the state and the economic system, it might function as an extension to the welfare state. But, the expectation of greater flexibility and efficiency of civil society in solving problems is an over-simplification. Rather, the increased active consideration of civil-social commitment in urban planning not only reveals that civil society does not merely consist of bottom-up movements that react to societal and political topics as independently corrective, but also proves governmental intervention in civil society. Therefore, top-down implementations, appearing notably in participation measures, need to be taken into account as part of the increased overlaps in the triangle.

Civil society's role can be analysed in the context of 'empty coffers' and the decline of state-run abilities as mentioned above, but the activation of commitment should be recognised as a form of valorisation well as. Both the valorisation of districts or neighbourhoods and the creation of communities are part of a competitive politics aimed at gaining a locational advantage and a satisfying position in 'successful cities' rankings. The logics of valorisation and rankings are inherent parts of neo-liberal governing,[1] since increasing competition for production, industry, and government funds drives cities to enhance their status and strengthen their potential by improving hard and soft locational factors (Heeg & Rosol, 2007; Schipper, 2013).

Analysing the activation of civil society in the neo-liberal city brings into focus the target groups of activation measures, the means and purposes of activating them, and the logics and techniques of activators. This chapter will give insights into the main theses of an ongoing research project on Berlin based on the relevant literature and empirical findings that substantiate the theoretical arguments by elaborating the new mode of governance.[2] It therefore outlines the genesis and operating mode of neighbourhood management, which serves as a field of activation, and illustrates how the new mode of regulation is established by raising civil commitment as a resource for governing and by focusing on communities. Furthermore, I argue, empowerment measures are combined with a 'responsibilisation' of citizens as well as a general valorisation of the third sector, which in turn result in a transformation of the conditions for participation and activism.

Berlin's neighbourhoods as fields of activation

In 1999, Germany's Red-Green coalition led by Chancellor Schröder started an urban development promotion programme between the State and the Federal States, called 'Neighbourhoods in special need of development – the Socially Integrative City'.[3] The programme's approach consisted in improving the quality of life in selected districts mainly via investments into urban development, but also by activating residents and coordinating cooperation between different stakeholders, such as politicians, business representatives, and residents. With a federal budget of €190 million in 2018 (Bundesministerium des Innern, für Bau und Heimat, 2018) it serves as the funding basis for neighbourhood management ('*Quartiersmanagement*'), which is the main implementation instrument in all 16 States.[4]

Neighbourhood management serves as a fruitful object of investigation in the context of participatory procedures and activation measures in order to consider the top-down implementation of public commitment as well as the interweaving of social spheres – economy, state, and third sector – in the city.

In Germany, neighbourhood management has already been practised for decades – in some areas that were particularly disadvantaged in order to avoid or attenuate social segregation even before the discourse around the Social City ('*Soziale Stadt*') began at the end of the 1980s (Güntner, 2007). One of its core debates was situated within urban sociology and dealt with poverty and segregation in cities as a spatial dimension of social exclusion (Häußermann, 2000), while the second debate was discussed mainly in political sciences and focused on the municipal welfare state's capacity to act (Hanesch, 1997). The debates intertwined around the question of possible policy recommendations to solve the crisis of the social welfare state at the regional level (Alisch, 1998;

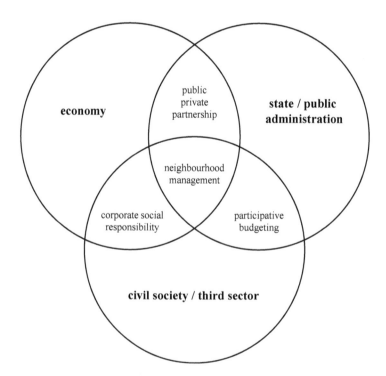

Figure 5.1 Schematic overlaps in the triangulation of societal spheres.

Gornig & Goebel, 2013; Güntner, 2007; Hanesch, 1997; Häußermann, 2000; Kronauer & Siebel, 2013) and provided the basis for the Socially Integrative City programme mentioned above.

Within this framework, neighbourhood management aims

> to stabilise neighbourhoods and strengthen their social cohesion (...) through the activation and intense participation of residents and local stakeholders, combined with investment in the neighbourhood's infrastructure.
> (Quartiersmanagement Berlin, 2018)

Local neighbourhood management offices serve as a 'contact point and communication centre' in individual neighbourhoods (Quartiersmanagement Berlin, 2018). With certain participation instruments, such as neighbourhood councils ('*Quartiersräte*')[5] and action fund juries ('*Aktionsfondjuries*'),[6] they want to improve the quality of life for residents in disadvantaged neighbourhoods, empower the articulation of needs, and thus allow for democratisation through the promotion of representative participation. Besides networking with the Senate department, initiatives, and other neighbourhood management offices, they organise public funds and flagship initiatives for improving districts' images. Moreover, they qualify, coach, and empower multipliers to establish a peer helper system in the spheres of education and health/exercise (details below).

One part of their partnership-based approach is the collaboration with housing associations. As written in the cooperation agreement between the Berlin Senate and the six municipal housing associations, a broad participation strategy is necessary to improve acceptance and quality of planning procedures (Senatsverwaltung für Stadtentwicklung und Wohnen, 2017a). Therefore, municipal housing associations increasingly focus on participatory measures as part of their 'social responsibility' (I4)[7] by aiming to activate residents, e.g. through tenant representations,[8] urban gardening projects, community centres, art workshops, and cooperation with the afore-mentioned neighbourhood councils. Participation and activation of tenants and investment in social projects are intended to improve and vitalise the district, to reduce crime rates and squalidness, which is then thought to lead to sustainable long-term tenant retention and stability in these neighbourhoods.

Both neighbourhood management offices and participation measures of municipal housing associations are located at the intersection of state, economy, and civil society. However, there are a lot of NGOs that share the narrative of activating civic action as well. They aim to influence socio-political discourses through campaigns and seek out funds to realise civil commitment and encourage residents to take on responsibility for the maintenance of public spaces. So, the mode of activating citizens comes to the fore, not only in neighbourhood management funded by a governmental programme or participatory measures of municipal housing associations as one example for municipal companies, but also in the practices of primary actors of civil society.

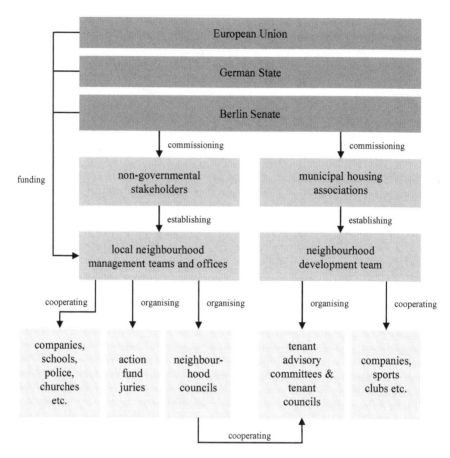

Figure 5.2 Institutional realisation of participatory measures.

This chapter focuses on Berlin, which is not only Germany's largest city, but also one of the cities that received the highest fundings from the Socially Integrative City programme between its initiation in 1999 and 2016 (Bundesinstitut für Bau-, Stadt- und Raumforschung, 2017a). Moreover, it is the city with the highest number of assisted areas.[9] Hence, Berlin suits the research topic outlined above because it offers many opportunities for a more detailed field investigation into the activation of civil society on the basis of neighbourhood management.

To address the thesis describing activating as an element of neo-liberal urban development processes, it is necessary to give at least a short classification of the city by highlighting two aspects. First, participatory planning procedures are important instruments for city authorities and local politicians in Berlin. In international comparison, several legacies of the Fordist welfare state still exist in this city, such as modest forms of rent control and tuition-free universities and, at least between 2001 and 2011, an intensified participative political culture (Lebuhn,

2015, p. 109; Kemp, Lebuhn, & Rattner, 2015, p. 708). Often-cited examples of the significance of activists who claim their rights include social movements against gentrification and privatisation as well as the referendum on the former Tempelhof airport, when the majority of Berlin's residents rejected politicians' and investors' plan and instead voted to keep the airfield as a huge park.

Second, in the previously divided city of Berlin that was shaped by capitalism in one part and socialism in the other, a 'late neo-liberalisation' can be identified (Lebuhn, 2015, p. 111). Nevertheless, if we take a look at more recent developments, there are other events that reveal neo-liberal characteristics as well. In particular, the so-called '*Berliner Bankenskandal*' and the associated financial crisis of 2001 resulted in drastic austerity measures (Lebuhn, 2015, p. 111). Privatisation, outsourcing, and public-private partnerships also changed 'Berlin's topography of formerly state-supplied goods and infrastructure' (Kemp, Lebuhn, & Rattner, 2015, p. 710). Therefore, the city's neo-liberal characteristics on the one hand and its participative political culture on the other allow an unprejudiced insight into activation modes which should not be categorised too hastily as either purely neo-liberal approach or democratic principle.

New modes of regulation

Civil commitment as a governing resource

Some decades ago, urban grassroots movements were demanding more participation, but the structures of local decision-making 'did not open up very far' (Mayer, 2003, p. 110). As 'politicians, urban scholars and activists in urban development now all highlight the importance of grassroots empowerment and citizen participation for dealing with urban problems', it seems likely that the claim has since been fulfilled (Mayer, 2003, p. 110). Meanwhile, however, many authors declare a new mode of governance that raises individual action as a governing resource. In general, the customary mode of urban governance has changed from planning using broad overviews to a regulation of more complex processes through moderating and involving relevant stakeholders. Thus, it is not simply a decline in regulation overall, but rather a different mode of regulation that fosters emancipation, one that interweaves planning and individual freedom (Kamleithner, 2009, p. 29).

This thesis refers to Michel Foucault's concept of governmentality, which he understood 'in the broad sense of techniques and procedures for directing human behaviour' (Foucault, 1994, p. 81). He was using the term for the 'deliberations, strategies, tactics and devices employed by authorities for making up and acting upon a population and its constituents to ensure good and avert ill' (Rose, 1996, p. 328). According to Foucault, governmentality can be defined as the 'ensemble formed by the institutions, procedures, analyses and reflections, the calculations and tactics, that allow the exercise of this very specific albeit complex form of power' (Foucault, 1991, p. 102). Following Rose, O'Malley, & Valverde (2006, p. 84), Foucault's concern was to analyse the birth of liberalism, which he understood as a political rationality, in particular as an art of governing that arose as a critique of excessive government. Neo-liberal critics problematised 'social government as generating government overload, fiscal crisis, dependency, and rigidity' and created another rationality for government 'in the name of freedom' through

> a range of techniques that would enable the state to divest itself of many of its obligations, devolving those to quasi-autonomous entities that would be governed at a distance by means of budgets, audits, standards, benchmarks, and other technologies that were both autonomizing and responsibilizing.
> (Rose, O'Malley, & Valverde, 2006, p. 91)

In neighbourhood management, we can recognise this kind of technology. A flyer for Quartiersmanagement Berlin (2018) for example states: 'the activation and involvement of the residents is one of the most important components of the neighbourhood work'. The official *Handbook of Participation* allows for a better understanding of how this demand is justified and how it should be realised. It was worked out on behalf of the then SPD-run Senate department to impart theoretical knowledge about participation and promote its usefulness. The handbook particularly addresses public administration staff and offers them a guide for involving residents and other important stakeholders in urban planning. Following the authors, Berlin's administration should turn towards moderating and coordinating partners of civil society, activating residents to get involved, and empowering them to help themselves (Walz et al., 2012, p. 39). This approach is justified on the one hand with the argument that participation is the foundation of democracy (Walz et al., 2012, p. 14). On the other hand, the increasing complexity and diversity of problems that have to be managed while public expenses are shrinking is another reason given (Walz et al., 2012, p. 37).

Here, the neo-liberally driven paradigm comes to the fore: in the handbook, we already find evidence for the thesis describing activation as a new mode of regulation using civil commitment as a governing resource. Urban governance is indeed required to establish new modes of regulation, since previously used modes apparently cannot fit with current performance expectations. Now, regulation modes orient progressively to the market as networks and urban administration must operate in an economically efficient manner and compensate shrinking budgets by outsourcing responsibilities to other stakeholders of civil society and private operators. The bottom line is that public administration is divesting itself of obligations by calling for greater participation from residents while simultaneously still wielding control through the moderation and coordination of activism. It is a matter of mobilising *and* regulating at the same time.

Inquiring into the activating logics and techniques, a new mode of regulation also appears in the framework conditions of neighbourhood management: the Berlin Senate[10] commissions private agencies and non-profit organisations to organise neighbourhood management in a specific area. Usually, the chosen organising institution runs an office in the particular assisted area. These areas are chosen based on indicators of the Monitoring for the Socially Integrative City Development[11] ('*Monitoring Soziale Stadtentwicklung*'), a surveillance programme observing the city's socio-spatial development since 1998, identifying areas at risk of being excluded from the city's development and, thus, in need of support from neighbourhood management to catch up (Senatsverwaltung für Stadtentwicklung und Wohnen, n.d.). Criteria include unemployment, long-term unemployment, social benefit payments, and child poverty (Senatsverwaltung für Stadtentwicklung und Wohnen, 2017b). In contrast, other dimensions of citizenship such as care (e.g. measurable by child care ratios) or dwelling (e.g. measurable by vacancy and fluctuation rates) are not considered. Currently, 34 neighbourhoods are supported by neighbourhood management with varying funding volumes (Quartiersmanagement Berlin, 2018). In general, the funding is temporary. The duration varies depending on the status of the particular area in comparison with other areas. Neighbourhood management is funded by the European Union, the German State and the respective Federal State based on the afore-mentioned Socially Integrative City programme.

The outlined procedure constructs areas as problematic and in need of tailored solutions. Urban administration is outsourcing development strategies by commissioning non-governmental stakeholders to work out specific local solution methods. In its 'Integrated Action and Development Concepts', stakeholders must explain how they will initiate and implement additional measures together with local partners and residents 'as an auxiliary to the municipal regulatory structures' (Quartiersmanagement Berlin, 2018). Afterwards, the commissioned stakeholders who run neighbourhood management offices have to prove their success on the basis of measurable results.[12] That way, urban development procedures are shifted from administrations to stakeholders who themselves acquire other local partners such as schools, NGOs or citizens' initiatives for the implementation of their strategic concepts.

The examples illustrate the new regulation mode of urban governance: on the one hand, the overall regulation is theoretically a duty of urban administration, as it selects areas in need and commissions organising institutions. On the other hand, however, with the aid of neighbourhood management, urban administration shifts responsibilities, especially the implementation of regulation tasks to local stakeholders who are supposed to coordinate the civil society actors. Here, the central assumptions of the above-mentioned *Handbook for Participation* guide the activities of neighbourhood management offices, too. Their activities focus on 'improving the quality of life of residents in a disadvantaged neighbourhood by activating them to take part in the development of their environment', as the head of a neighbourhood management office (I1) stated. So, planning and regulation is shifted to non-governmental stakeholders who empower civil commitment and foster emancipation. On behalf of the urban government, they coordinate local activities as intermediate actors and at the same time frame civil commitment as necessity for improved social cohesion and in times of low budgets.

'Governing through community' as a new rationality of governance

In this connection, human beings play an active role in their own self-government, analysed by Foucault in terms of technologies of the self. Their freedom is not opposed to government; rather it is one of the principal strategies of this specific kind of governmentality that Rose (1992) termed 'advanced liberal government' and that constructs freedom as the obligation to maximise one's life as a kind of enterprise. Following this argument, individuals 'come to understand and act upon themselves within certain regimes of authority and knowledge, and by means of certain techniques directed to self-improvement' (Rose, O'Malley, & Valverde, 2006, p. 90):

> Individuals are to be governed through their freedom, but neither as isolated atoms of classical political economy, nor as citizens of society, but as members of heterogeneous communities of allegiance, as 'community' emerges as a new way of conceptualizing and administering moral relations amongst persons.
>
> (Rose, 2009, p. 147)

Building on Foucault's concept of governmentality, Nikolas Rose is one of the authors who analysed the new rationality of governmental strategy as 'governing through community'. He points out 'the emergence of a range of rationalities and techniques that seek to govern without governing society, to govern through regulated choices made by discrete and autonomous actors in the context of their particular commitments to families and communities' (Rose, 1996, p. 328). His approach describes the intensified focus on quarters and communities as part of a new governance mode. These technologies 'involve (…) a variety of strategies for inventing and instrumentalizing (…) dimensions of allegiance between individuals and communities in the service of projects of regulation, reform or mobilization' (Rose, 1996, p. 334).

The new modes of neighbourhood participation, local empowerment, and engagement of residents in decisions affecting their own lives all aim to (re)activate self-motivation, self-responsibility, and self-reliance in the form of active citizenship within a self-governing community which 'is not simply the territory of government, but a means of government' (Rose, 1996, p. 335). Current urban governance regimes also exploit personal allegiances and active responsibilities. But, they do not focus only on communities in the narrow sense. Neighbourhoods, families, and individuals are important units, and the idea of state-centred welfare and anti-poverty policies is replaced by a 'human/social capital approach' (Kemp, Lebuhn, & Rattner, 2015, p. 706; Mayer, 2003). The concept of social capital plays a key role, as it connects 'local participation, based on horizontal networks and reciprocity, with such positive results as economic growth and democratic intensity' (Mayer 2003, p. 110).

An illustrating example for the human/social capital approach is the concept of so-called educational agents ('*Bildungsbotschafter*innen*'). Neighbourhood management offices aim to qualify and empower parents on various topics to make them 'better partners in education for kindergartens, schools, and community centres' and thereby improve educational opportunities for their children (I1). As soon as these educational agents are trained, they become peer helpers in schools, kindergartens, neighbourhoods and families, advising and guiding other parents. Special importance is given to the fact that they are able to communicate with their peer group in their mother tongue.

Here, we find an example of individuals who are called to work as connectors between the regulating unit – the neighbourhood management office as an agent of the Senate – and the specific target group, in this case educationally deprived social groups. It is not the educational establishment who is taking care of equal educational opportunities. The concept of educational agents suggests that peer helpers must be activated to allow fundamental possessions. The example not only proves the stated thesis that civil commitment becomes a governing resource as it enables the participation of marginalised groups. Additionally, it reveals that not only neighbourhoods (mentioned above as areas in need) are the focus of governing strategies. Communities that have to be reached through the commitment of qualified individuals are another target. These individuals serve as human or social capital in current regulation modes to realise participation and thus a democratic principle.

Reconfiguration of the third sector and transfer of responsibility

The above-mentioned examples highlight another aspect that must be examined: the 'reconfigurations in local state–society relations' which have an impact 'especially on the trajectory of third or voluntary sector development' (Mayer, 2003, p. 110). The encouragement of cooperation and networking of different stakeholders that guides Berlin's administration is not only helping to form a lean state. Empowerment measures also reflect a so-called soft neo-liberalism. Constructed local communities and protest movements are co-opted for the city's marketing and as guarantors of social cohesion in times of market-oriented governmental strategies. By building a sense of identity, citizens are urged to manage their lives and needs in an active and economically efficient manner (Rose, 2000). In addition to assuming this individual obligation, the status of the local level is generally enhanced and self-help organisations become cheap providers for services the nation-state cannot (or does not want to) supply anymore (Heeg & Rosol, 2007, p. 503). Therefore, an overall valorisation of the third sector and honorary services can be stated (Rose, 1996).

All interviewees agreed that social services and responsibilities formerly provided by the state or municipalities are transferred to local stakeholders. One of them explained, for example, that responsibilities of the Senate in the spheres of education, childcare, and health are shifted to neighbourhood management structures that are partially overstrained due to limited resources. The cooperation between the neighbourhood development team of a municipal housing association and a sports club represents the wide range of activities of neighbourhood development which is contested by the team itself. The team leader explained:

> If we cooperate with [the sports club] to reach overweight or even obese children in the neighbourhood, the question is – why? (…) Is it due to insufficient sport activities in schools, due to public health departments that failed to exert an influence on parents, as those who are actually responsible for the nutrition of their children? Sometimes I ask myself if this is really the task of a municipal housing association.
>
> (I4)

Although, theoretically, the Socially Integrative City programme only allows additional projects that do not belong to governmental tasks of the social welfare system, in fact it is always a matter of argument in the application for funds. A head of a neighbourhood management office stated: 'We do not take on the responsibilities of municipal administration. Otherwise, the particular project would not be eligible. We organise additional offerings. It is important to make that clear. Otherwise, we would not get the funding' (I1). It seems that the funding guidelines disguise a sort of transferred responsibility. Indeed, the regulatory framework for neighbourhood management seems ambivalent if we consider that educational and health projects are funded by a programme which was actually aimed at urban development investments. It can be assumed that the Socially Integrative City programme itself becomes a legitimising framework for shifting the welfare state's responsibilities vertically *and* horizontally.

Nevertheless, the (vertical) shift of responsibilities is not questioned by everyone. Stakeholders also argue that public administration cannot cover every aspect: 'Even a lot of money, a bunch of cleaning machines (…) cannot guarantee a clean city if its citizens do not become aware: me, myself, I have to do something against pollution', declared the representative of an NGO that promotes civic action in Berlin (I3). This quote exposes how responsibilisation is a mode of shifting tasks from state or municipalities not only to neighbourhood management structures, but also to every single citizen. Classical empowerment measures are often combined with a responsibilisation of citizens, even though it is not always obvious. Predominantly, the interviewees associated empowering people with emancipating them to 'articulate and enforce their claims', 'to awaken interest for their neighbourhood and to try to integrate them' (I1) – and quite often reaching those objectives is attempted by calling on everyone's duty as citizens who should play their part in improving neighbourhoods, urban and socio-spatial development, and finally their own quality of life.

This logic was already described in the concept of the activating society (*'Aktivgesellschaft'*), a term used by Lessenich (e.g. 2009, 2011) to describe the transformation of the welfare state in advanced capitalist societies. He defines the reconstruction of socio-political institutions as a reaction to the economic requirements of flexible capitalism (Lessenich, 2011, p. 254). In this regard, the

Governing through participation 89

activation of capitalist subjects serves as 'self-disciplination', drawing on a responsible lifestyle and intrinsic orientation towards market principles, but also as a programme of self-realisation by enabling participation and inclusion (Lessenich, 2011, p. 258). Very similarly, the ostensibly empowering measures in neighbourhood management and development tend to motivate citizens to engage voluntarily and play a part in compensating shrinking budgets. There are two main narratives underlying such justifications: one is that activated citizens secure their *own chances* for inclusion and general quality of life in a friendly, clean, and safe neighbourhood. The second narrative refers to advocating for the *common good* by serving the community, keeping an eye on the neighbourhood and engaging in public actions. Therefore, this kind of activation is referring both to citizens as responsible for themselves and for society. It is not necessarily and exclusively a part of neo-liberal governmentality: following Lessenich, the activating welfare state is also a 'neo-social' arrangement (Lessenich, 2011, p. 259). As such, it creates new dynamics of inclusion and exclusion in urban areas.

Transformation of conditions for participation and activism

New modes of regulation can be observed in the reconfiguration of local state–society relations as well as in a general responsibilisation. In addition, they result in new conditions for participation, involving processes of exclusion on the one hand and new room to manoeuvre for particular social groups on the other. Analysing the new opportunity structure for civic commitment, it is important to highlight these two sides of the coin by looking at the consequences, new forms of dependence, and ambivalences that result from this mode of activation.

The activation of public commitment is not only presented as an opportunity for participation as an active member of society. Rather, the invocation of civic action stresses the necessity of taking responsibility as a citizen. This new mode recodes activism, civil commitment, and citizenship that now 'has to be earned by certain types of conduct' (Rose, 2000, p. 98):

> This transformation *from citizenship as possession to citizenship as capacity* is embodied in the image of the active entrepreneurial citizen who seeks to maximize his or her own lifestyle through acts of choice, linked not so much into a homogeneous social field as into overlapping but incommensurate communities of allegiance and moral obligation.
> (Rose, 2000, p. 99, emphasis added)

Performance becomes an important factor for every citizen to guarantee his or her inclusion in society. So, do we even have to state a shift from a right to a duty in terms of social inclusion? We find evidence for this thesis by looking at the narratives surrounding the need for civil commitment, some of which were already mentioned above: the representative of an NGO that promotes civic action in Berlin stated, for example, that civic commitment should be a duty 'because one has to practise it' (I3). That way, it is constructed as a natural obligation of solidarity that must be internalised and exercised. Also, arguing that the collective cleaning of pavements and park areas serves as an important 'educational measure' (I3) underlines this obligational aspect. Moreover, a socially communicative aspect was emphasised by the same interviewee, explaining that civic action brings people together, as joint action avoids isolation by helping people to use public spaces more intensely instead of withdrawing into private spheres (I3). Another reasoning focuses on valorisation: the leader of the neighbourhood development team of a municipal housing association emphasised that engagement and identification help to revitalise public spaces, preventing discontentment and riots that result in vandalism, squalidness, and crime (I2). Empowering residents to take responsibility for themselves and their environment by using public offerings but also initiating their own projects (I2) is another central demand in the context of activation. In summary, most of the reasons underline the *necessity* of civic action in every society by calling on the natural duty of every individual. In the style of Kennedy's famous quote, the dominant claim would be: Ask not what your city's government can do for you – ask what you can do to improve your neighbourhood. Such an approach that is both autonomising and responsibilising has far-reaching consequences concerning the dynamics of inclusion and exclusion.

One argument is that the call for public commitment produces or intensifies social inequalities. While the social welfare state's strategies typically focus on disadvantaged social groups, the above-described new mode in neo-liberal cities might have the tendency to encourage already privileged groups (Kamleithner, 2009, p. 39). Although reaching marginalised social groups is one important guideline in neighbourhood management, it usually already fails at the language barrier due to limited resources. Further, opportunities to participate, such as the tenant representations mentioned above, are mostly noticed and taken by those with a high educational background who are able to articulate their demands in public anyway. Hence, exclusion is not suspended by measures that aim to engage all social groups.

At the same time, however, others tend to interpret the activation of commitment as an opportunity for a new form of subsidiarity in terms of an increasing chance for innovations due to new stakeholders who can extend the inner circle of current influencers and policymakers and therefore create pressure on decision-makers (Hanesch, 1997, pp. 9, 47–48, 50–51). For example, public forums are organised by neighbourhood management offices and conducted to facilitate the collaboration between tenants, administration, people in power, and other stakeholders in the neighbourhood, such as schools, companies, or the police. They aim to serve as an opportunity to raise one's voice and attract the municipal councillor's attention for local needs. Also, tenant representation in the form of advisory committees and tenant councils have meanwhile become a participatory requirement for municipal housing associations.

Governing through participation 91

The decision [for an additional tenant representation] was made by the Senate and we are just fulfilling it. We wouldn't have created such a dual structure. (...). A lot of foregoing considerations have not been made, such as: how can both tenant representations cooperate in a reasonable way?

(I4)

Mentioning some conflicts, the interviewee illustrates how an ostensibly participatory instrument demanded by the municipal state does not fit with reality and even produces dissatisfaction while failing to increase the chances for participation. In both examples, it remains debatable to what extent activating and participatory measures can actually cause changes in institutionalised political power structures and have an impact on decision-making processes in the political arena. To answer that question, further research is necessary.

Nevertheless, it must be considered that measures for activating commitment are not neo-liberally driven per se and intensify social exclusion as neighbourhood management even attempts to diminish neo-liberal consequences. An illustrative example is the conflicting relationship between neighbourhood development and gentrification: neighbourhood management offices find themselves handling the conflicting goals of developing certain areas in a positive way by initiating participatory projects to make neighbourhoods more attractive while at the same time avoiding the typical dynamics of capitalist valorisation processes. Improving the living environment of residents in a neighbourhood carries the inherent risk of displacement due to gentrification processes which in turn result in a termination of further neighbourhood management. Faced with the preconception of being the 'spearhead of gentrification' (I1), one neighbourhood management office decided: 'we initiated a project – we had to express that cleverly to secure its eligibility – it supports tenants or neighbourhoods that are affected by current developments of the housing market' (I1). The quote exposes the room for manoeuvre given to neighbourhood management institutions in their conflicting role between their mission, funding requirements, and social claims.

The balancing act is even intensified as neighbourhood management and development depend on sponsors who pursue their own interests by supporting civic action in order to cultivate their image. One interviewee mentioned an investor who tries to improve a bad image by sponsoring action days for cleaning the environment: '[They] try to compensate (...) and we fully accept that as a justified motivation' (I3). According to the representative of an NGO that promotes civic action in Berlin, another reason for cooperation seems to be to have some sort of control over social movements and possible protests: 'the BSR[13] has an explicit self-interest. (...) Actually, they try to coordinate initiatives, to support but in the broader sense also to control them'. (I3)

Besides those influences, other factors shape the opportunity structure for civic commitment as well. A lot of initiatives depend on the agreement of the urban administration due to the large number of approval procedures they have to pass through: based on the Monitoring for the Socially Integrative City

Development, the Senate department makes decisions about funding for areas in need. They choose the organising institutions for neighbourhood management and allow extensions or endings of the funding period. This gives the urban administration control and power, while shifting executive responsibilities to other stakeholders, such as neighbourhood management offices, neighbourhood development teams or non-governmental organisations.

Moreover, funding requirements seem to result in a reorganisation of civil society. Umbrella organisations become more important as necessary proposers in the application process for funding. This enables an even better bundling and thereby control of otherwise widespread initiatives, as one interviewee explained: 'citizens who want to get involved need an organisation that helps them to carry out their activism. But this organisation cannot rest exclusively upon volunteering', since the willingness towards civil commitment is changing – people want to act spontaneously, without obligation but with a clear impact (I3). Umbrella organisations seem to cater to these altered conditions for civic action. Urban administration also benefits, as this new way of regulating enables them to govern through ostensible empowerment measures.

Conclusion

This chapter argued for the phenomenon of activating civil society as part of a neo-liberal governmentality and a neo-social arrangement by pursuing the question: *Who is to be governed, and how?* Residents, tenants, and local stakeholders, such as business representatives, schools, companies, churches, police, etc., are target groups of activation measures. They are governed through top-down implementations, a new form of regulation which is also characterised as a transformation from 'government to governance'. This new mode of regulation increases individual action as a governing resource, as it fosters emancipation, while entwining planning and individual freedom. It rests upon a reconfiguration of local state–society relations and a general responsibilisation by calling residents and other stakeholders to take part in the development of their neighbourhood, for example, by establishing peer helper systems.

However, not only the state is managing the conduct of citizens, but rather a whole variety of authorities. Therefore, we need to ask: *Who governs whom?* It was shown that the top-down implementation between public administration and citizens is realised via organising institutions for neighbourhood management, neighbourhood development teams in municipal housing associations, and non-governmental umbrella organisations. Thereby, planning and regulation are shifted to non-governmental stakeholders, who activate and coordinate local activities on behalf of the urban government. Besides those intermediate actors, individuals become connectors between the regulating unit and the target group. The example of educational agents illustrated how they serve as human/social capital.

In order to determine *the logics* according to which the city is governed, different narratives were identified: the compensation argument refers to low budgets and a decline of state-run abilities, which necessitate an increased focus on civil society. In contrast, the cohesion argument focuses on improving the quality of life of residents in disadvantaged neighbourhoods by empowering them to articulate their needs and, thus, allow for democratisation through the promotion of representative participation. Besides the traditional narrative viewing volunteering as a part of social life, as a natural obligation of solidarity and a socially communicative act, moderating and involving partners in civil society aims to improve acceptance and quality in planning procedures. Additionally, the activation of civil commitment is justified as a valorisation of districts. In combination with the creation of communities, this is part of the competition argument: neighbourhoods are improved and vitalised due to lower crime rates, reduced squalidness, and social 'stability', which secures a solid position in inner-city rankings.

Further, the analysis focused on the *techniques of activation*. It was stated that regulation modes are oriented towards the market. Urban administration must operate in an economically efficient manner and therefore outsources responsibilities to other stakeholders of civil society and private operators. The new mode is characterised by mobilisation and regulation at the same time, becoming visible in the outsourcing of development strategies by commissioning non-governmental stakeholders. Simultaneously, public administration still has control via the moderation and coordination of activism, as it chooses the organising institutions, and defines requirements for municipal housing associations and areas in need. Through the activation of self-motivation, responsibility, and self-reliance, community becomes a means of government. Communities and protest movements are co-opted for the city's marketing as guarantors of social cohesion. Primarily, empowerment measures seem to enable a kind of autonomisation, but they are also combined with a responsibilisation of citizens who are called to be responsible for themselves and wider society.

What are the consequences of activating civil society? The new mode of regulation indicates a transformation of citizenship. The stated shift from right to duty in terms of social inclusion is based on an understanding of public commitment as a necessary responsibility. Rose's (2000, p. 99) diagnosis of a transformation 'from citizenship as possession to citizenship as capacity' puts it in a nutshell. The new mode shapes the opportunity structure for activism and civil society in urban areas in different ways: civic commitment depends on sponsors and funding requirements of urban administration, which thereby gains control over social movements. Moreover, new conditions for participation and activism create both exclusion and new rooms to manoeuvre for particular social groups. Activating measures might intensify social inequalities, as they mainly target already privileged groups. However, there could be an increased chance for innovations due to new stakeholders or those who use the full capacity of funding requirements to diminish neo-liberally driven processes in the city.

Further research is required to understand the conflicting relationship between the responsibilising mode of activation and inclusion or exclusion through the perspective of citizens. It can be assumed that participatory instruments convince citizens, as they promise inclusion and self-fulfilment by drawing on individual freedoms. Nevertheless, some might also tend to refuse activation. The effects on citizens who do not behave according to the described rationality of governance still need to be elaborated.

Notes

1 See sociology of evaluation (e.g. Lamont, 2012; Vatin, 2013).
2 The research project follows a 'pragmatic grounded theory' approach (Timonen, Foley & Conlon 2018), remaining open to unanticipated findings even though knowledge of the literature shapes the conceptual design and entry into the field. The current data base consists of semi-structured interviews, concepts available online and website contents of a neighbourhood management office, a neighbourhood development team of a municipal housing association, and a non-governmental organisation that promotes civic action in Berlin. All of them focus on different spheres of activity, which ensures a certain degree of heterogeneity in the data base and extends the understanding of how the general conditions shape social reality.
3 The programme '*Stadtteile mit besonderem Entwicklungsbedarf – die soziale Stadt*' is also translated as 'City Districts with Special Development Needs – the Social City' (Quartiersmanagement Berlin, 2018). However, based on Kemp, Lebuhn, and Rattner (2015), in this chapter the translation 'the Socially Integrative City' will be used below as a short form of the programme name.
4 Until 2017, about 850 assisted areas in 450 cities and municipalities received funding from the Socially Integrative City programme (Bundesministerium des Innern, für Bau und Heimat, 2018). In 2005 85 per cent of the areas registered possessed at least one neighbourhood management office (Bundesinstitut für Bau-, Stadt- und Raumforschung, 2017b, pp. 33, 53).
5 Neighbourhood councils represent the concerns of all the residents in a neighbourhood, irrespective of the property owners. They consist of representatives of the residents and other stakeholders, such as headteachers, representatives of religious communities, sports clubs, neighbourhood initiatives. Members of neighbourhood councils decide how to use the money from the neighbourhood fund for projects in that neighbourhood.
6 Action fund juries decide about the funding of short-term activities and projects up to €1,500 organised by residents and stakeholders to activate and support the neighbourhood.
7 All interviews (marked as I1/I2/I3/I4) were conducted in German. Quoted passages were translated into English and are not literal renditions in the strict sense.
8 Tenant advisory committees ('*Mieterbeiräte*') were selected in Berlin already in the 1980s. The elected members of the advisory committee are intermediaries for the tenants of the neighbourhood and represent their concerns towards the housing association. Since the new 'Berliner Wohnraumversorgungsgesetz' in 2016, tenant councils ('*Mieterräte*') are another requirement that municipal housing associations have to fulfil. Tenant councils function as a committee for all tenants of a housing association and are involved in corporate planning and neighbourhood development
9 Data for 2014: Berlin: 41, Hamburg 18, Bremen 12 assisted areas (Bundesinstitut für Bau-, Stadt- und Raumforschung, 2017b, p. 33).
10 More specifically: Senatsverwaltung für Stadtentwicklung und Wohnen.
11 Also translated as 'Urban Social Development Monitoring' (Quartiersmanagement Berlin, 2018).
12 Such as attendance at events, full spending of funds, number of people volunteering in the neighbourhood.
13 BSR: Berliner Stadtreinigungsbetriebe, waste management company of Berlin.

References

Alisch, M. (Ed.) (1998). *Stadtteilmanagement: Voraussetzungen und Chancen für die Soziale Stadt*. Opladen: Leske + Budrich.

Brenner, N. (2004). *New state spaces: Urban governance and the rescaling of statehood*. Oxford and New York, NY: Oxford University Press.

Brenner, N., & Theodore, N. (2002). Cities and the geographies of 'actually existing neoliberalism'. *Antipode, 34*(3), 349–379.

Bundesinstitut für Bau-, Stadt- und Raumforschung. (2017a). Bundesfinanzhilfen je Stadt/Gemeinde im Programm Soziale Stadt 1999 bis 2016 in Euro. Retrieved from www.bbsr.bund.de/BBSR/DE/Stadtentwicklung/Staedtebaufoerderung/Forschungsprogramme/SozialeStadt/KartenGraphiken/karte-2016-suvol.pdf?__blob=publicationFile&v=6.

Bundesinstitut für Bau-, Stadt-, und Raumforschung. (Ed.) (2017b). *Zwischenevaluierung des Städtebauförderungsprogramms Soziale Stadt*. Bonn: Bundesinstitut für Bau-, Stadt- und Raumforschung.

Bundesministerium des Innern, für Bau und Heimat (2018). Soziale Stadt. Retrieved from www.bmi.bund.de/DE/themen/bauen-wohnen/stadt-wohnen/staedtebau/soziale-stadt/soziale-stadt-node.html.

Evans, B., Richmond, T., & Shields, J. (2005). Structuring neoliberal governance: The nonprofit sector, emerging new modes of control and the marketisation of service delivery. *Policy and Society, 24*(1), 73–97.

Foucault, M. (1991). Governmentality. In G. Burchell, C. Gordon, & P. Miller. (Eds.) *The Foucault effect: Studies in governmentality: With two lectures by and an interview with Michel Foucault* (pp. 87–104). Chicago, IL: The University of Chicago Press.

Foucault, M. (1994). *Ethics. Subjectivity and truth. The essential works of Foucault, 1954–1984*. New York, NY: New Press.

Gornig, M., & Goebel, J. (2013). Ökonomischer Strukturwandel und Polarisierungstendenzen in deutschen Stadtregionen. In M. Kronauer, & W. Siebel (Eds.), *Polarisierte Städte: Soziale Ungleichheit als Herausforderung für die Stadtpolitik* (pp. 52–69). Frankfurt and New York, NY: Campus.

Güntner, S. (2007). *Soziale Stadtpolitik: Institutionen, Netzwerke und Diskurse in der Politikgestaltung*. Urban Studies. Bielefeld: Transcript.

Hanesch, W. (Ed.) (1997). *Überlebt die soziale Stadt? Konzeption, Krise und Perspektiven kommunaler Sozialstaatlichkeit*. Opladen: Leske + Budrich.

Häußermann, H. (2000). Die Krise der 'sozialen Stadt'. *Aus Politik Und Zeitgeschichte, 10/11*, 13–21.

Heeg, S., & Rosol, M. (2007). Neoliberale Stadtpolitik im globalen Kontext. Ein Überblick. *PROKLA – Zeitschrift für kritische Sozialwissenschaft, 37*, 491–509.

Hess, S., & Lebuhn, H. (2014). Politiken der Bürgerschaft. Zur Forschungsdebatte um Migration, Stadt und Citizenship. *sub\urban. Zeitschrift Für Kritische Stadtforschung, 2*(3), 11–34.

Kamleithner, C. (2009). 'Regieren durch Community': Neoliberale Formen der Stadtplanung. In M. Drilling, & O. Schnur (Eds.), *Governance der Quartiersentwicklung* (pp. 29–47). Wiesbaden: VS.

Kemp, A., Lebuhn, H., & Rattner, G. (2015). Between neoliberal governance and the right to the city: Participatory politics in Berlin and Tel Aviv. *International Journal of Urban and Regional Research, 39*(4), 704–725.

Kronauer, M., & Siebel, W. (Eds.) (2013). *Polarisierte Städte: Soziale Ungleichheit als Herausforderung für die Stadtpolitik*. Frankfurt and New York, NY: Campus.

Lamont, M. (2012). Towards a comparative sociology of valuation and evaluation. *Annual Review of Sociology, 38*(1), 201–221.

Lebuhn, H. (2015). Neoliberalization in post-wall Berlin. Understanding the city through crisis. *Critical Planning. UCLA Journal of Urban Planning, 22*, 98–119.

Lessenich, S. (2009). Mobilität und Kontrolle. Zur Dialektik der Aktivgesellschaft. In K. Dörre, H. Rosa, & S. Lessenich (Eds.), Soziologie – *Kapitalismus – Kritik: Eine Debatte* (pp. 126–177). Frankfurt a.M.: Suhrkamp.

Lessenich, S. (2011). Die kulturellen Widersprüche der Aktivgesellschaft. In C. Koppetsch (Ed.), *Nachrichten aus den Innenwelten des Kapitalismus* (pp. 253–263). Wiesbaden: VS.

Mayer, M. (2003). The onward sweep of social capital: Causes and consequences for understanding cities, communities and urban movements. *International Journal of Urban and Regional Research, 27*(1), 110–132.

Quartiersmanagement Berlin (2018). Information on the Berlin neighbourhood management programme. The social city programme. Retrieved from www.quartiersmanagement-berlin.de/fileadmin/user_upload/SSE_Einlegeblatt_2S_2016_engl-WEB.pdf.

Rose, N. (1992). *Towards a critical sociology of freedom: An inaugural lecture given by Professor Nikolas Rose.* London: Goldsmiths College.

Rose, N. (1996). The death of the social? Re-figuring the territory of government. *Economy and Society, 25*(3), 327–356.

Rose, N. (2000). Governing cities, governing citizens. In E. F. Isin (Ed.), *Democracy, Citizenship, and the Global City* (pp. 95–109). London and New York, NY: Routledge.

Rose, N. (2009). Governing 'advanced' liberal democracies. In A. Sharma, & A. Gupta (Eds.), *The anthropology of the state: A reader* (pp. 144–162). Blackwell Publishing.

Rose, N., O'Malley, P., & Valverde, M. (2006). Governmentality. *Annual Review of Law and Social Science, 2* (1), 83–104.

Schipper, S. (2013). *Genealogie und Gegenwart der 'Unternehmerischen Stadt': Neoliberales Regieren in Frankfurt am Main 1960–2010.* Münster: Westfälisches Dampfboot.

Senatsverwaltung für Stadtentwicklung und Wohnen (2017a). Leistbare Mieten, Wohnungsneubau und soziale Wohnraumversorgung. Kooperationsvereinbarung mit den städtischen Wohnungsbaugesellschaften Berlins. Retrieved from www.stadtentwicklung.berlin.de/wohnen/wohnraum/wohnungsbaugesellschaften/download/kooperationsvereinbarung.pdf.

Senatsverwaltung für Stadtentwicklung und Wohnen (Ed.) (2017b). Monitoring Soziale Stadtentwicklung Berlin 2017. Kurzfassung. Retrieved from www.stadtentwicklung.berlin.de/planen/basisdaten_stadtentwicklung/monitoring/download/2017/MSS2017_Kurzfassung.pdf.

Senatsverwaltung für Stadtentwicklung und Wohnen (n.d.). Monitoring Soziale Stadtentwicklung. Retrieved from www.stadtentwicklung.berlin.de/planen/basisdaten_stadtentwicklung/monitoring/.

Swyngedouw, E. (2005). Governance innovation and the citizen: The janus face of governance-beyond-the-state. *Urban Studies, 42*(11), 1991–2006.

Timonen, V., Foley, G., & Conlon, C. (2018). Challenges when using grounded theory: A pragmatic introduction to doing GT research. *International Journal of Qualitative Methods, 17*(1), 1–10.

Vatin, F. (2013). Valuation as evaluating and valorizing. *Valuation Studies, 1*(1), 31–50.

Walz, S., Kast, A., Schulze, G., Born, L., Krüger, K., & Niggemeier, K. (2012). *Handbuch zur Partizipation.* Berlin: Kulturbuch.

Part III
Everyday experience of urban neo-liberalisation

6 Permanent liminality?

Housing insecurity and home

Hannah Wolf

Introduction

A specific figuration of success and crisis can be witnessed in nearly every contemporary metropolis, widely known as *The Housing Crisis*. In general, the term is invoked to refer to multiple interconnected phenomena: it is used to point out the general shortage of affordable housing for ever-growing parts of the urban population (Eurostat, 2016); to describe low or sub-standard living conditions in existing housing provision; to refer to the decline of public or social housing that can be observed in the past decades (Aalbers & Holm, 2008); to connote the rising numbers of evictions and foreclosures (EACRHC 2016); ultimately, to refer to the growing number of homeless people living on the street as well as so-called hidden homeless who are sofa-surfing, living in hostels, or temporary accommodation and who tend to not appear in official homelessness statistics (Soederberg, 2018). Against this backdrop it is obvious that the specific characteristics of *The Housing Crisis* vary between countries, cities, and even boroughs. There is not one housing crisis but, in fact, multiple housing crises.

To be sure, housing is not in crisis for everyone, and the invocation of the term *housing crisis* in public debate might very well be veiling the fact that there is also a housing success for actors who are making profit. A general, albeit short caveat is to be made regarding the term crisis as such, referring to Reinhart Koselleck (1979). Going back to Greek origins of the word, a crisis foremostly denotes a turning point; the verb κρίνειν means 'to decide', and it is in crisis that decisions are made – for better or worse:

> It is the essence of crisis that a decision is due, but not yet made. The openness of the decision is part of the crisis. The general uncertainty of a critical situation is accompanied by the certainty that (…) the end of the critical stage is near. The possible solution remains unclear, yet the end itself, a turning over of the existing conditions – threatening and dreaded or hopefully desired – is guaranteed.
>
> (Koselleck, 1979, p. 105, translation H. W.)

In other words, crisis is both a situation of 'irreducible contingency' (Makropoulos, 2013, p. 15) and of enforced decision-making. Koselleck acknowledges that it is the usage as an *economic* term which has turned crisis into a central concept of modernity. Here, throughout the eighteenth and nineteenth century in liberal as well as in socialist thought, progress is thought to be achieved through decisions made in reoccurring times of crisis. Decisions are seen as inevitable: to diagnose a crisis is at the same time to demand measures to end it,

> although the determination of the optimal time for a decision is now thought to be determined by inescapable pressures for action. At that moment, use of the concept of crisis is meant to reduce the room for manoeuver, forcing the actors to choose between diametrically opposed alternatives.
>
> (Koselleck, 2006, p. 370)

In this vein, Madden and Marcuse point out that the invocation of the term *The Housing Crisis* can serve as an 'ideological distortion' (Madden & Marcuse, 2016, p. 10), supporting seemingly simple solutions such as deregulation of planning and building guidelines, resulting in a steady transfer of power to developers – justified and legitimised by the diagnosis of a crisis and the political pressure to end it as fast as possible. Within a capitalist political economy, however, where housing functions as a commodity, a housing crisis turns out to be not a contingent, but a *necessary* part of capitalist cycles of accumulation (Harvey, 1985). The occurrence of a housing crisis is very much the opposite of an irregularity or illness, and one needs to be careful not to neutralise underlying systemic forces and power imbalances at play by reproducing a certain diagnosis and thereby justifying hastily made decisions (Marcuse, 1988). One current example of how political and economic elites are employing *The Housing Crisis* is by diagnosing a lack of housing and proposing ways to increase building activity, as is the case in Germany. Another example is the diagnosis of people's inability to afford to buy houses and the proposition to design new financial instruments to subsidise property ownership, i.e. through equity loans such as *Help to Buy* in the UK. In these cases, the diagnosis of a crisis and the subsequently proposed solutions follow an arguably simple economic model of supply and demand. They define a problem – *not enough houses* or *not enough buyers* – and aim to solve it by incentivising growth of either buildings or capital. The structural forces that might have caused the crisis in the first place are rendered irrelevant by the 'inescapable pressures for action'. In these cases, however, the solutions appear to resemble mere band-aids.

Against this backdrop, I propose to view *The Housing Crisis* as a 'transitional phase made necessary by the immanent logic of a prior development' (Richter & Richter, 2006, p. 356). Drawing on ongoing qualitative research about private renting in London and Berlin, this chapter aims at elucidating the connections and contradictions between wider political-economic transformations, their consequences on the built environment at local-urban level, and, ultimately, their implications for the lives of people, their lived experiences and their meanings of home. There are two main reasons for choosing the private rental sector (PRS) in these two cities: first, we are witnessing a transformation in the tenure structure in both cities; with a shift from private renting to homeownership in Berlin (Heeg, 2013), and vice versa in London (Ronald & Kadi, 2018). Second, in both cities, renting has become associated with a number of social problems: the PRS in the UK is one of the main factors driving inequality, poverty, and homelessness, especially in London where rent is on average more than twice compared with the rest of the country and the number of private renters living in poverty has more than doubled in the past decade (Trust for London, 2017). Repossessions of rented homes by landlords are at an all-time high, and accordingly the number of people living in temporary accommodation and on the streets has been steadily increasing (GLA, 2017, p. 65). Berlin, albeit traditionally being the 'renters' capital' – and Germany the single EU country with nearly even numbers of owner-occupation and renting – has experienced a thorough destabilisation of its formerly affordable PRS. Average rent prices have risen by 71 per cent from 2010 to 2017 (Investitionsbank Berlin, 2018), contributing to Berlin's current status as the city with the fastest growing property prices (Collinson, 2018). Although the city administration does not collect reliable data concerning the connections between the housing market, evictions and homelessness (Soederberg, 2018), recent research suggests that

> [the] changed dynamics of Berlin's housing market have caused a new quality of housing emergencies. Terminations of rental contracts, eviction notices and forced evictions have become regular instruments to increase profits.
>
> (Berner, Holm, & Jensen, 2015, p. 93)

How does the transformation of the private rental sector affect renters' experience and meaning of home? In order to pursue this question, this chapter first provides a theorisation of housing as a multi-faceted phenomenon (Ruonavaara, 2018), second presents the trajectories that have produced the current renting insecurity in London and Berlin, and third investigates private renters' experiences of home. The insecurity of renting both in London and Berlin, I argue, can be conceptualised as permanency of a state of 'betwixt and between' (Turner, 1969, p. 107) that not only challenges peoples' capabilities to make a home and feel at home, but also points towards a possible transformation of shared meanings of home.

Theorising housing

Housing is situated at the core of socio-spatial relations: relations of power and dependency literally become built into the urban environment and form the material infrastructure of the lives of individuals and communities; rights of property and tenure situate people within power relations; and housing as home provides the basis for a range of socio-spatial relations, practices, and routines as well as being 'a place where the soul can rest' (hooks, 2009, p. 143). Analytically, three ideal-typical meanings of housing in contemporary capitalist societies can be distinguished: housing as *commodity*, housing as a *right*, and housing as *home*.

Housing as *commodity* refers to the economic dimension of houses as objects that are produced, consumed, and traded in the market. Like all commodities, it has a dual character expressed as exchange value and use value. With a focus on exchange value, housing becomes real estate: something that can be built, bought, sold, invested in, and speculated on. Housing's use value comes into focus as dwelling or the 'use of housing' (King, 2004): housing provides protection and shelter, a space for social reproduction, and individual and subjective fulfilment of interests (Ruonavaara, 2018). Following Marx, these types of value stand in close dialectic tension; the owner of a house can realise its exchange value only by surrendering its use value to someone else (Harvey, 2015), e.g. by selling it or renting it out.

Housing as a *right* addresses the recognition of the right to housing as stated in the *Universal Declaration of Human Rights* and the *International Covenant on Economic, Social and Cultural Rights*. In this respect, it refers to relations between states and their citizens, since it ultimately remains the obligation of state institutions to implement and ensure the fulfilment of human rights (Nash, 2009). Housing thus becomes an issue of citizenship: of inclusionary and exclusionary practices regarding distribution of and access and entitlement to housing.

Finally, housing as *home* reflects a whole range of socio-spatial and psychosocial aspects. Home can be understood both as place and feeling, a nodal point of various social relations, experience of continuity through space and time, as well as expression of identity (Douglas, 1991; Duyvendak, 2011; Easthope, 2004; Mallett, 2004; Porteous & Smith, 2001; Rybczynski, 1986; Somerville, 1992). Nearly all authors struggle with a comprehensive definition of the concept of home, due to its many layers of meaning ranging from intimate and domestic to national to global scale, and from subjective emotions and experiences to culturally contingent ideological and political usage (Moore, 2000). Duyvendak points out that it is the very familiarity of home in everyday life and language that contributes to the elusiveness of home as a concept:

> while everyone agrees that we know what it is to feel at home, the moment we have to describe what it means to us, we begin to stutter. Feeling at home, then, is one of those emotions that eludes words.
> (Duyvendak, 2011, p. 27)

It is, however, he concludes 'in a *non-essentialist* sense – an *essential*, even existential, feeling for all' (Duyvendak, 2011, p. 27): a feeling that is at once both universal *and* radically subjective. Thus, analytically, home is as much a dimension of housing as housing is a dimension of home, understood as practice and experience of rootedness, belonging, and 'ontological security' (Giddens, 1991).

In everyday language, housing and home are often conflated; but home can be more than housing, and housing can be more – and less – than home. Indeed, the multifacetedness and multiscalarity of home need to be made productive:

'the home, the neighbourhood and the nation are all potential spaces of belonging [and] each of these spaces conditions the other' (Morley, 2001, p. 433, cited in Duyvendak, 2011, p. 4). Thus home, albeit frequently being associated with and sometimes indeed epitomising the private realm, is always a 'public issue', too (Mills, 1959). Against this background I propose to understand the dimensions of commodity and right as opportunity structures (Merton, 1995; Mackert, 2010) of home. Opportunity structures differ depending on individual social positioning; they facilitate and restrict agency and the realisation of life-chances. Home in this context is understood as a fundamental life chance, providing the basis for the 'socio-emotional security systems of the self' (Dunn, 2013, p. 188). This basis, I argue, is becoming fragile as there are ever more urban inhabitants in London and Berlin.

Producing renting insecurity through urban neo-liberalism

The transformation of opportunity structures of housing within past decades can be analysed through the lens of 'urban neo-liberalism' (Brenner & Theodore, 2002), a concept that has emerged out of neo-liberalism's contested character as ideology and process: while neo-liberal *ideology* assumes that market forces operate 'according to immutable laws no matter where they are "unleashed"' (Brenner & Theodore, 2002, p. 349), analyses of neo-liberalisation *processes* show 'actually existing neo-liberalism' as a variegated and contextually embedded phenomenon. Cities have become 'increasingly central to the reproduction, mutation, and continual reconstitution of neo-liberalism' (Brenner & Theodore, 2002, p. 375) as sites of both strategic and experimental implementation of neo-liberal policies as well as of protest, contestation, and resistance (Mayer, 2013). As an analytical tool, urban neo-liberalism aims at the description, comparison, and critique of the manifold realisations and consequences of global political-economic decisions and local restructurations. Despite contextually contingent differences in quality and intensity, these neo-liberalisation processes are generally characterised by a 'progressive dismissal of the public sector' (Vanolo, 2017, p. 67) through the prioritisation of 'market-based, market-oriented, or market-disciplinary responses to regulatory problems, (...) commodification in all realms of social life (and the mobilisation of) speculative financial instruments to open up new arenas for capitalist profit-making' (Brenner, Peck, & Theodore, 2010, pp. 329–330). The 1993 World Bank report is a case in point for the neo-liberalisation of housing. Under the title *Housing: Enabling markets to work* it compiles a set of strategies to advise governments' housing policies, including the development of private property rights and mortgage finance and the reduction of housing subsidies by explicitly advocating against public housing and rent control (World Bank, 1993, p. 46).

Actually existing neo-liberalism comprises dynamic processes of privatisation, deregulation, and financialisation. *Privatisation* is generally understood as the transfer of services from the public to the private sector. From a sociological point of view, it needs to be regarded relationally within a wider context of welfare restructuring, such as the erosion of social rights and ensuing social inequalities (Turner, 2016); furthermore, as a hegemonic project proceeding through ideological means and conveying particular 'truths' that serve to legitimise the restructuring that is taking place (Samson, 1994). *Deregulation* refers to all those processes in which the state retreats from control and intervention in social and economic relations: the well-known 'roll back' movement in order to liberate markets and facilitate market behaviour in any realm (Theodore, Peck, & Brenner, 2011). However, deregulation is not simply a process of retreat; it entails *re*-regulatory processes and policies that aim at establishing rules and boundaries that flank and facilitate supposedly free and unregulated market behaviour (Jessop, 2013). Finally, *financialisation* refers to processes in which financial actors and logics penetrate into non-financial sectors, and in which 'the entire urban ensemble and the practices of participants have become (…) both a *product* of financialised accumulation and a primary *instrument* enabling that accumulation to take place' (Edwards, 2016, p. 28). While it is possible to distinguish these processes analytically, it makes no sense to view them as distinct from one another. As parts of a general logic of neo-liberalisation they reinforce and empirically co-constitute each other. In the realm of housing, together they form the 'normal, mundane operations of the world's [and the city's] political economy' (Porteous & Smith, 2001, p. 106). The housing trajectories of London and Berlin serve to illustrate the systematic production of renting insecurity through these processes.

London

For London, 8 August 1980 marks the beginning of housing privatisation: part of the Housing Act under the Conservative government led by Margaret Thatcher was the *Right to Buy* scheme (RTB) that offered public housing units for sale to sitting tenants at great discount prices while at the same time preventing reinvestments into social housing. RTB was the bedrock of the vision of a 'property-owning democracy' (Thatcher, 1975) that aimed at fostering 'attitudes of independence and self-reliance' through the naturalisation and normalisation of a 'desire for homeownership' (House of Commons, 1980). In effect, home-ownership rates rose from 49.2 per cent in 1971 to 64.1 per cent in 1987 with the privatisation of two million housing units altogether until to date. Up to the 1980s, living in social housing was common with one in three UK households renting from local authorities; today, social housing is associated with high levels of economic and social deprivation and the stigmatisation of tenants in so-called 'sink estates' (Slater, 2018). Altogether, 40 per cent of London's public housing stock was transformed into private property. While there is no doubt that those who could afford to buy their homes did benefit from the scheme, RTB also produced growing inequalities: council housing waiting lists keep growing and more and more people become reliant on finding accommodation in the PRS. Furthermore, a large number of RTB properties can be found in today's PRS: three years after the sale, more than 25 per cent of RTB properties in inner London were not inhabited by the original buyers, but rented out privately (Jones, 2003; Murie, 2016); recent research estimates that across London more than 36 per cent of RTB properties are let by private landlords (Copley, 2014). Effectively, RTB has turned into *Right to Let* and led to the emergence of two polarised classes, 'generation rent' and 'generation landlord' (Ronald & Kadi, 2018). While 'generation rent' represents individuals and households who are unable to afford homeownership due to rising real estate prices, stagnating incomes, and reduced social benefits, and thus depend on renting privately, 'small-time landlords' (Ronald & Kadi, 2018, p. 791) are an equally growing part of London's population for whom renting out is a form of individual asset-based welfare and security against future risk and poverty in later life (Smith, Albanese, & Truder, 2008). Due to the persistent scarcity of social housing, the number of households receiving housing benefit in the PRS is growing, effectively transferring 'over 40 per cent of the entire housing benefit (...) into the pockets of private landlords' (Minton, 2017, pp. 92–93). In order to reduce public expenditure, a number of welfare reforms have since limited the amount of housing benefit, largely regardless of whether the amounts received suffice to pay private rents (Clarke, Hamilton, Jones, & Muir, 2017).

The abolishment of rent control is a clear example of deregulation in London. Rent control and 'regulated tenancies' in the PRS had been a vital part of the UK housing system since 1915 (Kemp, 2015). During the 1980s, these measures were defined as housing *problems* by the Conservative government, claiming they were the cause for poor dwelling conditions through lack of investment. This claim was in tune with a general criticism of state regulation and intervention and an emphasis on private enterprise and progress. The Housing Acts of 1988 and 1996 abolished all forms of private rent regulation and re-regulated the PRS through the introduction of *assured shorthold tenancies* (AST) as default tenancy, with renting contracts regularly set at a fixed duration of six (or 12) months, thereby thoroughly reducing security of tenure for private renters. After the assured period, landlords can evict tenants with two months' notice under Section 21 of the 1988 Housing Act. There is evidence that this legislation, infamously known as 'no fault eviction', is used by landlords and letting agents to increase rents, by either evicting tenants and renting out to new ones at a higher price, or by informally threatening with eviction in order to pressurise tenants into more expensive contracts. Over 60 per cent of nationwide uses of Section 21 take place in London where the repossession rate has risen by 87 per cent from 2003 to 2015 (Clarke, Hamilton, Jones, & Muir, 2017, p. 15). These numbers only reflect the documented uses of Section 21 as a legal instrument; they do not show in how many cases it is used as an informal instrument to exert power and pressure on tenants. Research findings suggest that tenants will put up with higher costs, inadequate dwelling conditions, and conflicts with landlords simply out of fear that they might otherwise lose their home (Smith, Albanese, & Truder, 2014; Age UK, 2017; Rhodes & Rugg, 2018). Furthermore, Section 21 has been brought in connection with the phenomenon of 'revenge evictions' after tenants have complained or requested repairs (Smith, Albanese, & Truder, 2014). For renters, ASTs ensue 'recurrent displacement' (Watt, 2018), and rising rents make finding an affordable place to live virtually impossible for low-income households. Since 2009, the number of people reporting homeless to councils has been steadily rising; 39 per cent of these homelessness cases result from terminated ASTs, and Londoners represent the majority of households in temporary accommodation (Trust for London, 2018).

The creation of *Buy to Let* mortgages (BTL) as new financial instruments played a significant role in the growth of the PRS: these mortgages were offered to private landlords first in 1996, shortly after the World Bank Report, at a relatively high rate of interest and consequentially did not account for a significant share of the mortgage market. This changed in the early 2000s when transformations of international capital markets – the internationalisation of finance, securitisation, and a low-interest-rate regime – resulted in relatively cheap and easy to secure mortgages specifically designed for private landlords. The almost guaranteed capital gains associated with BTL under conditions of rising property prices attracted novice landlords who were focused on capital growth and used their investment in housing as a means not only of accumulation and high returns, but also as a form of pension saving in the context of an asset- or property-based welfare system that socialises risks and shifts welfare responsibilities to individuals (Kemp, 2015, p. 609; Ronald & Kadi, 2018, p. 789). While BTL facilitated short-term investment behaviour on the side of private landlords, it also resulted in growing insecurity on the side of private tenants since a common condition of BTL mortgages is that they oblige mortgagors to grant tenancies for a maximum of one year's duration. Restrictions regarding letting to recipients of housing benefit are similarly a widespread feature of BTL loans, known as 'DSS discrimination' (Shelter, 2018). As a result, low-income households are severely struggling to find or keep a place to live in London. It is important to note here that the implementation of these discriminatory practices and policies follows an instrumental logic of financial securitisation which is facilitated by the interdependencies of the relationship between individual tenant and landlord and the one between local and international markets of capital. Housing in this respect is understood solely as speculative investment, be it as a vehicle for capital returns in the global market or as an alternative form of social insurance.

London's neo-liberalisation of housing started with a huge wave of privatisation, accompanied by deregulation processes in the 1980s which transformed the city's tenancy structure. Rather than increasing homeownership rates, however, these measures resulted in the emergence of a highly unregulated and growing PRS in which both private landlords' and renters' lives are connected to global markets through financial instruments within a context of welfare state erosion, impacting on individual rationalities with the effect that 'generation landlord' view their housing as an asset rather than a place to live, while the right to housing has been severely eroded and conditionalised for 'generation rent'.

Berlin

For Berlin, the decision to abolish the common public interest legislation of housing (*Wohngemeinnützigkeit*) in 1990 marks the beginning of a deregulation process on a national scale. Although security of tenure and tenancy rights remained relatively strong and stable, this decision changed the structural conditions of housing: the legislation that had been in place since 1940 compelled housing companies to a non-profit policy in exchange for tax deductions and thus secured moderate rent prices. After 1990, housing companies became 'normal' players in the housing market, logics shifted towards profit-orientation with the possibility to get listed on the stock exchange, and rents started rising disproportionately.

The first big wave of privatisation in Berlin, however, was due to developments on the municipal level: attempts to consolidate the city's severely indebted household led to a large-scale privatisation of social housing (Lebuhn, 2015). While Berlin owned 19 housing companies and 28 per cent of all housing units in 1991, the construction and maintenance of public housing came to a complete halt from 1995 onwards, gradually selling over 200,000 flats to private housing companies and reducing the public housing stock to 15.8 per cent by 2008. Privatisation happened largely en-bloc through sales to institutional investors who were able and eager to invest large amounts of capital. Thus, individual homeownership rates remained stable and low (12 per cent in 2008), while for many renters their tenure status gradually changed from public to private.[1] The city tried to sell as much and as quick as possible, additionally disposing of two of its housing companies and the sale of several housing estates to private investors. Berlin's housing privatisation subsequently paved the way for an uneven urban development resulting from roughly two ideal-typical investment rationalities – value-added vis-à-vis opportunistic strategies: value-added strategies targeted housing with 'development potential' (Uffer, 2013, p. 159) over a longer period of time, typically good quality housing estates in the inner city that were modernised and thus gained use value for affluent households; while opportunistic strategies aimed at short-time profits via the rapid buying and re-selling of property with no 'development potential' within volatile market and credit cycles, typically focusing exclusively on the property's exchange value and reducing management and maintenance costs to an absolute minimum in order to extract higher profits from lower rental payments (Uffer, 2013, p. 166). Thus, Berlin's housing privatisation widened socio-spatial inequalities and can be seen as a catalyst of gentrification and displacement of lower-income households. The proclaimed aims of privatisation – household consolidation and modernisation of the existing housing stock – were only partially accomplished and it is now politically recognised that the privatisation did not benefit the city as a whole; however, attempts to turn back this development and to remunicipalise privatised housing are faced with financial and legislative obstacles.[2] Additionally, a slow but persistent trend towards homeownership is observable in Germany with Berlin no exception: since 2009 the numbers of condominium conversions and individual purchases have been steadily rising (Investitionsbank Berlin, 2018), which can be explained by – in international comparison – relatively affordable property prices and the prospect of long-term value increase on the one hand, and the reduction of social security on the other, encouraging investments in so-called 'concrete gold' as a form of individual self-provision (Heeg, 2013). State measures to promote homeownership through subsidies (such as *Wohn-Riester* and *Baukindergeld*) also play a role, albeit a smaller one than in London.

Berlin's housing financialisation is shaped by institutional investors and their involvement in global markets where housing is an asset for trade and speculation, and a vehicle for capital increase or 'capital switching' (Harvey 1985). With this exclusive focus on housing as a bearer of exchange value, its use value becomes relevant only in cases in which it represents a means of increasing exchange value, e.g. through modernisation or luxurisation. Real estate funds illustrate another aspect: here, individuals become both subjects and objects of investments in the property market through real estate fund shares. As responsibilised investment subjects, people buy real estate shares as part of private asset-based welfare provision (Heeg, 2013); as investment objects, the flats that they live in are bought by international investment funds, such as the Danish private pension fund PFA that acquired 3700 German housing units, amongst them 400 flats in Berlin. Lastly, the purchase of condominiums follows a financialised logic as an investment with guaranteed profit, be it as a secure residence for the future, security against inflation, or private pension saving.

Neo-liberalisation of housing in Berlin started with national deregulation, followed by a wave of privatisation on the city scale, in the course of which rental housing became an attractive object for institutional and individual investors. This development was embedded in deregulation processes in the global financial market, the dismantling of the welfare state and the promotion of individual property ownership. The consequential segregation and gentrification processes are partially met by the city's 'social conservation measures' (*Milieuschutz*); however, these measures have limited effects. Against the backdrop of rising rents and scarce availability, the right to housing enshrined in the Berlin constitution can be regarded as at least contested (Krennerich, 2018).

Home under conditions of neo-liberalised housing

In both London and Berlin, rental housing has become a vital part of strategic asset development, with a widening gap between those who can and those who cannot afford individual purchase or investment (Helbrecht & Geilenkeuser, 2012). For the latter, the different trajectories and contexts of housing neo-liberalisation have led to respective typical situations: in London, private renters face a *general insecurity of tenure* due to the institutionalisation of ASTs and the possibility of getting evicted via Section 21 notices. *Recurrent displacement* is thus a second general feature of London's housing experience, capturing the phenomenon of repeated *involuntary residential mobility* (Hartman, Keating, & LeGates, 1982; Watt, 2018), as well as restricted chances of finding an affordable place to live. The rising number of people living in temporary accommodation and the insecurity associated with private renting in a context of *impossibility of homeownership* point towards a *perpetuity of urban housing precarity* (Wacquant, 2008; Watt, 2018). In Berlin, renters also encounter *insecurity of tenure*, mainly due to the possibility of changing ownership, the ongoing trend of condominium conversions and the ensuing risk of rent increase or termination of rental contracts through repossession notices (Fields & Uffer, 2016). While renters' rights were not formally restricted and rental contracts are generally indefinite, *displacement pressure* due to rent rises is common, with little chance to find new affordable housing. Private renting is thus currently undergoing a transformation from a formerly 'taken-for-granted' secure tenancy to a more *precarious and unpredictable tenure status*.

How do private renters experience home under conditions of neo-liberalised housing? One answer is provided by Madden and Marcuse (2016) with the terms 'residential oppression, exploitation and alienation'. In their interpretation, home can no longer provide protection against outside forces and instead the private realm becomes the arena of social violence, renters fall victim to powerful actors and interests. Similarly, Porteous and Smith employ the term 'domicide' to refer to the 'deliberate destruction of home' by outside forces (Porteous & Smith, 2001, p. 3; critically: Nowicki, 2014). However, people are not only victims of uncontrollable forces, but also active agents in the production of their circumstances, and home as human practice points towards the malleable character of this production: 'home is made, unmade and remade across the life course, subject to a seemingly unending variety of factors' (Nowicki, 2014, p. 788). In what ways do the changing structural conditions in the PRS imprint on the relationship between people and what they call their home and in what ways do people negotiate and *do home* under conditions of neo-liberalised housing? To pursue this question, the remainder of this chapter draws on ongoing qualitative research with private renters in London and Berlin involving narrative, biographic interviews, participant observation, media and social media analysis, and interviews with social service representatives. This research is grounded in the conviction that home is an active human practice akin to what Giddens (1991) has termed 'ontological security'. Making a home and making oneself at home are universal human practices responding to the need to experience a 'sense of continuity and order in events' (Giddens, 1991, p. 243). As a practice, home has both spatial and temporal dimensions and is also connected to self-identity as it provides the basis for a 'colonisation of the future' (Giddens, 1991, p. 182): for actively developing ideas and making plans for the future, for 'creating future territories of possibilities' (Giddens, 1991, p. 242). As Giddens argues, such reflexive projects of the self depend on a sense of basic trust, a taken-for-grantedness of certain features of the world and the 'bracketing-in' of risk and existential questions. In situations of housing insecurity, this taken-for-grantedness becomes fragile, an experience that can be conceptualised as *liminality* (Van Gennep, 1909). Liminality denotes a situation of equally *no more and not yet*, and can be employed to analyse individual as well as collective transitory and transformative periods:

> [Something] that happens in real life, whether for an individual, a group, or an entire civilization, that suddenly questions and even cancels previously taken-for-granted certainties, thus forcing people (…) to reflect upon their experiences, even their entire life, potentially changing not only their conduct of life but their identity.
>
> (Szakolczai, 2009, p. 158)

Describing private renting as liminal makes sense in two ways: first, it can be argued that collectively taken-for-granted assumptions about the distribution of and access to housing have not only become fragile, but that especially the PRS in London represents an 'incorporation and reproduction of liminality into "structures"' (Thomassen, 2009, p. 17). Second, these structural transformations lead to liminal situations in individual experiences of housing, e.g. when loss of housing via eviction or termination is anticipated. The analysis of these liminal situations allows for reflecting the changing nature of those practices and emotions that constitute the home.

Existential insecurity

While interview partners in Berlin stress the moment that they realised their insecure housing situation as shocking and existentially threatening, for London renters these feelings have become part of their everyday experience of renting. Londoners refer to a general insecurity with regard to their actual housing situation as well as to their status and prospects within the PRS: 'about to move knowing that we will be moving again in the new year'.[3] They describe being 'stuck' in private renting with neither social housing nor home-ownership a viable alternative. For Berlin renters, too, a growing mistrust towards the PRS can be observed: 'There is this thought, if we move now, who can guarantee that the same won't happen with the next flat?' This insecurity is spilling over to other aspects of life; a couple whose flat was sold and who were told by the new owner that they are planning repossession 'probably within the next two years' stress how this is affecting their sense of future: 'we really don't know (…) if you have no idea where you are going to end up you can't even apply for kindergarten'. Experiencing liminality as a spatial in-betweenness – *no longer here but not yet somewhere else* – becomes obvious in both cities; likewise, temporal in-betweenness is stressed through the permanency of waiting, an experience that can be captured by the term 'displacement anxiety': 'a *prospective* ruptured sense of place – of home and/or neighbourhood – as a result of a potential, forced external real-world move' (Watt, 2018 p. 74).

Emotional aspects of displacement anxiety

The emotional experience of displacement has been analysed as a form of grief (Fried, 1963). The emotions evoked through *prospective* or *anticipated* displacement encompass fear, anger, sadness, and resignation. In Berlin, interviewees talk about fear mostly with regard to their insecure future – when will I have to move, where can I move to, what happens if I don't find a place? The intensity of this emotion correlates with socio-economic position: fear is greater for low-income households with precarious labour conditions and little social security. However, fear is not only associated with the current housing situation, but turns into general anxiety towards the future: 'This is so depressing, now as a young person, if you don't have a well-paid job at 20 and can buy something, you will never have the chance (…) it's thinking about the future that depresses me'.

In London, the connection between fear and home is strongly articulated. On the one hand there is great fear of landlords' actions – 'you are unable to create a home as constant fear of being sold or rent increases and fear of eviction' – on the other hand, there is also general resignation: 'it's too late for my generation but the future is bleak'. In addition to fear, anger and aggression are the most frequently expressed emotions that are directed against certain groups of people – in London landlords and letting agencies, in Berlin buyers and investors – as well as against 'politics' in general. Renters describe two different strategies to cope with their anger: resistance/protest and repression/compensation. Resistance encompasses both subtle actions such as postponing viewing appointments or unfriendly behaviour towards potential buyers as well as collective protest actions by tenants' alliances. However, the difficulties of collective action are emphasised – protest requires resources, and many renters ask themselves if mobilisation is worthwhile: 'that would be wasted energy because I don't have any opportunities to do anything to defend myself'. Coping strategies range from compensation through 'more sport' to complete repression, as expressed by a former tenant who was forcefully evicted from her flat: 'I just did not believe that they would do that. I thought if I just carried on as usual, I could keep the flat'.

Violation of privacy

Several interviewees discuss the private/public divide, interpreting their insecure housing situation as a clear violation of privacy. While in Berlin viewing appointments are mostly referred to as intrusions into the private sphere, in London a restriction of privacy is seen as a general condition of private renting:

> I really detest the letters we receive saying they will let themselves in for their far too regular property inspections, where they take unnecessary photographs of your possessions (…). This is just not a home as owners constantly violate your privacy.

The necessity of privacy to make a home is clearly articulated, and the invasion into the private sphere is perceived as a violation of the integrity of home and potentially one's identity. Feelings of embarrassment and shame are shared by several interviewees. One Berlin tenant reports of a viewing:

> This young man from Austria said that he felt uncomfortable being in my private rooms while I was still there. I looked at him and said, well, I am really sorry, I feel uncomfortable all the time with strangers running around in my private rooms and not knowing what is going to happen. So, to me this was so presumptuous, rude, just impossible.

The feeling of embarrassment and shame by both parties refers to the shared perception of a transgression; at the same time the tenant's perspective emphasises the inequality of social status (Neckel, 1991) between her and 'this young man from Austria'. The reported rejection and self-assertion stress how *home* is put into position against *commodity*: the potential buyer is 'in my private rooms' – the possessive pronoun clearly marks a symbolic claim to ownership beyond legal ownership title.

Power and agency

Both in Berlin and London, tenants report feeling powerless and inferior, specifically with regard to having to move involuntarily. Realising that having a home depends on decisions of others is consistently described as an experience of powerlessness, and forced residential moves are interpreted as a loss of individual agency. The power imbalances between renters and landlords are emphasised by one Berlin tenant:

> You get the impression that they have different rights, different rules. They have the money, and it just works for them. That makes me aggressive because I can't do anything, it doesn't matter whether I have a lawyer, whether we solidarise, in the end you can't do anything.

This experience contradicts ideas of home as a place of personal control, independence, and protection from social constraints (Mallett, 2004, p. 70). In London, renters regularly feel at the mercy of more powerful actors and institutions: landlords raising rents or terminating contracts, the political and legal system privileging ownership rights and cutting housing allowances, and market dynamics that are believed to make matters still worse. These power dynamics are also inscribed into the home-space: 'home starts by bringing some space under control' (Douglas, 1991, p. 289), and it is exactly the lack of control over space that London renters describe, e.g. by being prohibited from renovating an apartment, removing owners' furniture or even hanging their own pictures on the walls: 'it's like living in a show home, only all the furniture is faded and shredded'.

Moral: injustice and guilt

Finally, insecurity of home is articulated very clearly and strongly as injustice, mostly addressed at the state. As one Berlin tenant expresses: 'The flats are for people who live in the state and pay taxes, and that the state doesn't protect our flat and says this will definitely not be a speculation object, that's not right'. Here, the dimensions of housing as commodity and as right are perceived as mutually exclusive, accompanied by the unsettling realisation 'that our rights are taken away from us'. While on the one hand there is a strong claim of political accountability and responsibility, it is noteworthy that there is also a dimension of personal guilt: 'Somehow I feel guilty about the flat. I feel guilty for my weakness (...) and I have this feeling, well, you should have been faster and more active'. The sense of guilt stems from assuming responsibility for a situation (Neckel, 1991) and can be interpreted as part of governmental subjectivation processes in which individuals are responsibilised to 'practice self-care' (Foucault, 1988; Rose, 1996). This is associated with a compulsion to greater flexibility in everyday life: 'It would be too abstract to look for something new now (...) I don't even know where I am going to work next year (...) so, you wait and look and have to stay flexible'. Social constraints are thus re-coded into self-constraints, the insecure housing situation is seen as partly due to personal failure, and individual behavioural change the only solution.

This double attribution of responsibility towards politics and oneself is observable in London, too. Private tenants describe on the one hand feeling neglected and forgotten within a 'nation of homeowners', on the other hand they express great disappointment about not being able to fulfil their goals of home-ownership. Buying a home represents a major milestone in what is seen as British *normal biographies*, and 'renting sits just above homelessness and marks a graded boundary of social exclusion' (Ronald, 2008, p. 10). The impossibility of homeownership is thus perceived as a personal failure to create a secure home and associated with being a 'second-class citizen'. Asked about what would happen if they were still renting at the age of 35, one London tenant responds: 'I'd probably feel like a bit of a failure in a way. I think my parents would be asking the question of why aren't you a home owner?' (McKee & Soaita, 2018, p. 16).

Conclusion

This chapter argued that in order to make sense of the various current housing crises they need to be understood as transitional periods resulting from long-term global and locally contextualised neo-liberalisation processes that thoroughly affect the lives and minds of people. In London and Berlin, a general transformation of housing structure is taking place, foremostly affecting the tenure of private renting: while in London the primacy and privileging of individual property rights led to the emergence of a highly unregulated PRS, in Berlin deregulation and the gaining domination of institutional investors shifted housing logics towards profit-orientation, resulting in growing insecurity of hitherto stable housing conditions in the PRS. These dynamics are embedded in a general dismantling of the welfare state and accompanying responsibilisation of citizens: while in London 'small-time landlordism' (Ronald & Kadi, 2018) provides opportunities for individual property-based welfare provision, in Berlin renters become subjects and objects of investments within a financialisation of housing (Heeg, 2013). Ideologies of housing and individual responsibility become especially relevant in the British context, illustrated by the phrase 'property-owning democracy' and the association of citizens' freedom and responsibility through homeownership (Nowicki, 2018); such associations are not observable to the same extent in the German context – however, assessing how promotion of private housing property is framed ideologically promises fruitful (Wolf, 2018). For both cities, a general shift can be stated: housing as a commodity has gained greater significance, while housing as a substantial right is becoming limited and conditionalised.

These transformations also affect private renters' experiences and meanings of home. Conceptualising renting insecurity as a form of structural and experiential liminality (Szakolczai, 2009), the qualitative analysis revealed five dimensions in which liminality of home becomes relevant: existential, emotional, private/public, power/agency, and injustice/guilt. These dimensions delineate the symbolic arenas in which private renters perceive threats to the integrity of home and in which they renegotiate their meaning of home. While the insecurity is objectively and subjectively higher and liminal to a larger degree in London than in Berlin, in both cities the meaning of home is negotiated and restructured within certain contradictory fields: between home as a place of individual agency and control vs. inferiority and powerlessness; between home as a place that enables a colonisation of the future vs. a place of 'displacement anxiety'; finally, between home as something that one has to afford vs. home as something that one – as citizen, as human – deserves. In this context, further attention needs to be paid to accounts of personal failure, shame, and guilt, since they point towards a success of neo-liberalised subjectivation processes that foremostly rely on individual responsibility (Rose, 1996). At the same time, it becomes apparent that private renters do not simply give up the idea of home. Both in London and Berlin renters' movements have emerged, claiming the right to housing and the right to the city and aiming to change existing structures of property, market, and power. Still: most tenants experience renting insecurity on an individual, personal level. Depending on their situation in the life course, their socio-economic resources, their embeddedness in social contexts and the intensity of this liminal experience, this will lead to variegated individual actions and consequences. With 85 per cent private renters in Berlin and an estimated 60 per cent by 2025 in London, the collective liminal experience holds the potential to transform shared meanings of home – in what way is yet to be seen.

Notes

1 This gradual shift is due to the structure of the social housing provision in Germany, organised as a public subsidy of any kind of housing providers in exchange for the use of a dwelling for social purposes on a temporary basis. Public subsidies decrease progressively and at the same time the rent increases until, after a period between 12 and 40 years of amortisation, the housing unit can be sold or rented out at market rates.
2 The initiative *Deutsche Wohnen enteignen* is currently aiming at employing political and legal instruments for the city to 'buy back' privatised housing stock (see www.dwenteignen.de/).
3 All quotes, unless otherwise referenced, are taken from original data (interviews and social media research the author conducted since March 2018).

References

Aalbers, M. B., & Holm, A. (2008). Privatising social housing in Europe. The cases of Amsterdam and Berlin. In K. Adelhof, B. Glock, J. Lossau, & M. Schulz (Eds.), *Urban trends in Berlin and Amsterdam* (pp. 12–23). Berlin: Geographisches Institut der Humboldt-Universität zu Berlin.

Age UK (2017). *'Living in fear'. Experiences of older private-renters in London.* London: Age UK.

Berner, L., Holm, A., & Jensen, I. (2015). *Zwangsräumungen und die Krise des Hilfesystems. Eine Fallstudie in Berlin.* Berlin: Humboldt-Universität zu Berlin.

Brenner, N., & Theodore, N. (2002). Cities and the geographies of 'actually existing neoliberalism'. *Antipode, 34*(3), 349–379.

Brenner, N., Peck, J., & Theodore, N. (2010). After neoliberalization? *Globalizations, 7*(3), 327–345.

Clarke, A., Hamilton, C., Jones, M., & Muir, K. (2017). *Poverty, evictions and forced moves.* York: Joseph Rowntree Foundation.

Collinson, P. (2018). Berlin tops the world as city with the fastest rising property prices. *Guardian*, 10 April. Retrieved from www.theguardian.com/world/2018/apr/10/berlin-world-fastest-rising-property-prices.

Copley, T. (2014). *From right to buy to buy to let.* London: Greater London Authority/London Assembly Labour.

Douglas, M. (1991). The idea of home: A kind of space. *Social Research, 58*(1), 287–307.

Dunn, J. R. (2013). Security, meaning, and the home: Conceptualizing multiscalar resilience in a neoliberal era. In P. Hall & M. Lamont (Eds.), *Social resilience in the neoliberal era* (pp. 183–205). Cambridge: Cambridge University Press.

Duyvendak, J. W. (2011). *The politics of home. Belonging and nostalgia in Western Europe and the United States.* London: Palgrave Macmillan.

EACRHC – European Action Coalition for the Right to Housing and the City (2016). *Evictions across Europe.* Retrieved from https://housingnotprofit.org/files/EvictionsAcrossEurope.pdf.

Easthope, H. (2004). A place called home. *Housing, Theory and Society, 21*(3), 128–138.

Edwards, M. (2016). The housing crisis: Too difficult or a great opportunity? *Soundings. A Journal of Politics and Culture 62*, 23–42.

Eurostat. (2016). *Urban Europe. Statistics on cities, towns and suburbs.* 2016 Edition. Luxembourg: Publications Office of the European Union.

Fields, D. J., & Uffer, S. (2016). The financialisation of rental housing: A comparative analysis of New York City and Berlin. *Urban Studies, 53*(7), 1486–1502.

Fried, M. (1963). Grieving for a lost home: Psychological costs of relocation. In J. Q. Wilson (Ed.), *Urban renewal: The record and the controversy* (pp. 359–379), Cambridge & London: M.I.T. Press.

Foucault, M. (1988). Technologies of the self. In L. H. Martin, H. Gutman, & P. H. Hutton (Eds.), *Technologies of the self: A seminar with Michel Foucault* (pp. 16–49). Amherst: The University of Massachusetts Press.

Giddens, A. (1991). *Modernity and self-identity: Self and society in the late modern age.* Stanford: Stanford University Press.

GLA – Greater London Authority (2017). *Housing in London: 2017.* London: GLA.

Hartman, C., Keating, D., & LeGates, R. (1982): *Displacement: How to fight it.* Berkeley, CA: National Housing Law Project.

Harvey, D. (1985). *The urbanization of capital: Studies in the history and theory of capitalist urbanization.* Baltimore, MD: Johns Hopkins.

Harvey, D. (2015). *Seventeen contradictions and the end of capitalism.* London: Verso.

Heeg, S. (2013). Wohnungen als Finanzanlage: Auswirkungen von Responsibilisierung und Finanzialisierung im Bereich des Wohnens. *sub\urban. zeitschrift für kritische stadtforschung, 1,* 75–99.

Helbrecht, I., & Geilenkeuser, T. (2012). Demographischer Wandel, Generationeneffekte und Wohnungsmarktentwicklung: Wohneigentum als Altersvorsorge? *Raumforschung und Raumordnung, 70*(5), 425–436.

hooks, b. (2009). *Belonging: A culture of place.* New York, NY and London: Routledge.

House of Commons (1980). Housing bill. Order for second reading, 15 January 1980. Retrieved from https://api.parliament.uk/historic-hansard/commons/1980/jan/15/housing-bill#column_1445.

Investitionsbank Berlin (2018). *IBB Wohnungsmarktbericht 2017.* Berlin: IBB.

Jessop, B. (2013). Putting neoliberalism in its time and place. *Social Anthropology, 21*(1), 65–74.

Jones, C. (2003). *Exploitation of the right to buy scheme by companies.* London: Office of the Deputy Prime Minister.

Kemp, P. A. (2015). Private renting after the global financial crisis. *Housing Studies, 30*(4), 601–620.

King, P. (2004). *Private dwelling. Contemplating the use of housing.* London: Routledge.

Koselleck, R. (1979). *Kritik und Krise: Eine Studie zur Pathogenese der bürgerlichen Welt.* Frankfurt a.M.: Suhrkamp.

Koselleck, R. (2006). Crisis. Translated by M. W. Richter. *Journal of the History of Ideas, 67*(2), 357–400.

Krennerich, M. (2018). Ein Recht auf (menschenwürdiges) Wohnen? *Aus Politik und Zeitgeschichte, 68*(25–26), 9–14.

Lebuhn, H. (2015). Neoliberalization in Post-Wall Berlin. Understanding the City through Crisis. *Critical Planning, 22,* 99–119.

Mackert, J. (2010). Opportunitätsstrukturen und Lebenschancen. *Berliner Journal für Soziologie, 20*(4), 401–420.

Madden, D., & Marcuse, P. (2016). *In defense of housing. The politics of crisis.* London: Verso.

Makropoulos, M. (2013). Über den Begriff der Krise. *INDES Zeitschrift für Politik und Gesellschaft, 1,* 13–20.

Mallett, S. (2004). Understanding home: A critical review of the literature. *The Sociological Review, 52*(1), 62–89.

Marcuse, P. (1988). Neutralizing homelessness. *Socialist Review*, *88*(1), 69–97.
Mayer, M. (2013). Urbane soziale Bewegungen in der neoliberalisierenden Stadt. *sub\ urban. zeitschrift für kritische stadtforschung*, *1*, 155–168.
McKee, K., & Soaita, A. M. (2018). *The 'frustrated' housing aspirations of generation rent*. Glasgow: UK Collaborative Centre for Housing Evidence.
Merton, R. K. (1995). Opportunity structure: The emergence, diffusion, and differentiation of a sociological concept, 1930–1950s. In F. Adler & W. S. Laufer (Eds.), *The legacy of anomie theory. Advances in criminological theory* (pp. 3–87). New Brunswick: Transaction Publishers.
Mills, C. W. (1959). *The sociological imagination*. New York, NY: Oxford University Press.
Minton, A. (2017). *Big capital: Who is London for?* London: Penguin.
Moore, J. (2000). Placing *home* in context. *Journal of Environmental Psychology*, *20*(3), 207–217.
Morley, D. (2001). Belongings: Place, space and identity in a mediated world. *European Journal of Cultural Studies*, *4*(4), 425–448.
Murie, A. (2016). *The right to buy? Selling off public and social housing*. Bristol: Policy Press.
Nash, K. (2009). Between citizenship and human rights. *Sociology*, *43*(6), 1067–1083.
Neckel, S. (1991). *Status und Scham: Zur symbolischen Reproduktion sozialer Ungleichheit*. Frankfurt am and New York, NY: Campus.
Nowicki, M. (2014). Rethinking domicide: Towards an expanded critical geography of home. *Geography Compass*, *8*(11), 785–795.
Nowicki, M. (2018): A Britain that everyone is proud to call home? The bedroom tax, political rhetoric and home unmaking in U.K. housing policy. *Social & Cultural Geography*, *19*(5), 647–667.
Porteous, J. D., & Smith, S. E. (2001). *Domicide: The global destruction of home*. Montreal, Kingston, London and Ithaca, NY: McGill-Queens University Press.
Rhodes, D., & Rugg, J. (2018). *Vulnerability amongst low-income households in the private rented sector in England*. York: University of York.
Richter, M., & Richter M. W. (2006). Introduction: Translation of Reinhart Koselleck's 'Krise' in Geschichtliche Grundbegriffe. *Journal of the History of Ideas*, *67*(2), 343–356.
Ronald, R. (2008). *The ideology of home-ownership: Homeowner societies and the role of housing*. London: Palgrave Macmillan.
Ronald, R., & Kadi, J. (2018). The revival of private landlords in Britain's post-homeownership society. *New Political Economy*, *23*(6), 786–803.
Rose, N. (1996). The death of the social? Re-figuring the territory of government. *Economy and Society*, *25*(3), 327–356.
Ruonavaara, H. (2018). Theory of housing, from housing, about housing. *Housing, Theory and Society*, *35*(2), 178–192.
Rybczynski, W. (1986). *Home: A short history of an idea*. New York, NY: Penguin.
Samson, C. (1994). The three faces of privatisation. *Sociology*, *28*(1), 79–97.
Shelter (2018). *Ending DSS discrimination*. Retrieved from https://blog.shelter.org.uk/2018/08/ending-dss-discrimination/.
Slater, T. (2018). The invention of the 'sink estate': Consequential categorisation and the UK housing crisis. *The Sociological Review*, *66*(4), 877–897.
Smith, M., Albanese, F., & Truder, J. (2014). *A roof over my head: The final report of the sustain project*. Shelter and Crisis. Retrieved from http://england.shelter.org.uk/__data/assets/pdf_file/0005/760514/6424_Sustain_Final_Report_for_web.pdf.

Smith, S. J., Searle, B. A., & Cook, N. (2008). Rethinking the risks of home ownership. *Journal of Social Policy, 38*(1), 83–102.

Soederberg, S. (2018). The rental housing question: Exploitation, eviction and erasures. *Geoforum, 89,* 114–123.

Somerville, P. (1992). Homelessness and the meaning of home: Rooflessness or rootlessness? *International Journal or Urban and Regional Research, 16*(4), 529–539.

Szakolczai, A. (2009). Liminality and experience: Structuring transitory situations and transformative events. *International Political Anthropology, 2*(1), 141–172.

Thatcher, M. (1975). *Speech to conservative party conference on 10 October 1975.* Retrieved from www.margaretthatcher.org/document/102777.

Theodore, N., Peck, J. & Brenner, N. (2011). Neoliberal urbanism: Cities and the rule of markets. In S. Watson & G. Bridge (Eds.), *The New Blackwell Companion to the City* (pp. 15–25). Oxford: Wiley.

Thomassen, B. (2009). The meaning and uses of liminality. *International Political Anthropology, 2*(1), 5–28.

Trust for London (2017). *London's poverty profile 2017.* Retrieved from www.trustforlondon.org.uk/publications/londons-poverty-profile-2017/.

Trust for London (2018). *Repossessions and evictions.* Retrieved from www.trustforlondon.org.uk/data/repossessions-and-evictions/.

Turner, B. S. (2016). We are all denizens now: On the erosion of citizenship. *Citizenship Studies, 20*(6–7), 679–692.

Turner, V. (1969). *The ritual process.* New York, NY: Aldine de Gruyter.

Uffer, S. (2013). The uneven development of Berlin's housing provision. In M. Bernt, B. Grell, & A. Holm (Eds.), *The Berlin reader. A compendium on urban change and activism* (pp. 155–170). Bielefeld: transcript.

Van Gennep, A. [1909] (1960). *The rites of passage.* London: Routledge.

Vanolo, R. (2017). *City branding: The ghostly politics of representation in globalising cities.* London: Routledge.

Wacquant, L. (2008). *Urban Outcasts: A comparative sociology of advanced marginality.* Cambridge: Polity Press.

Watt, P. (2018). 'This pain of moving, moving, moving': Evictions, displacement and logics of expulsion in London. *L'Année Sociologique, 68*(1), 67–100.

Wolf, H. (2018). Debatte Baukindergeld: Nur ideologische Kosmetik. *die tageszeitung,* 13 July. Retrieved from www.taz.de/!5517410/.

World Bank (1993). *Housing: Enabling markets to work. A World Bank policy paper.* Washington: World Bank. Retrieved from http://documents.worldbank.org/curated/en/387041468345854972/Housing-enabling-markets-to-work.

7 Athens in times of crisis

Experiences in the maelstrom of EU restructuring

Dina Vaiou

Introduction

Ten years into the recent multi-faceted crisis, developments in the European Union are marked by deepening socio-spatial inequalities, dismantling of welfare policies, and growing levels of poverty, accompanied by rising support for the discourse and practices of political parties of the far right. It is by now well-documented that the economically stronger countries have deferred much of the necessary adaptations to the weaker economies, mainly but not exclusively in the European south (see, among many, Euro Memo Group, 2019). Greece is one of the countries which have borne the destructive effects of such practices, through the implementation of extreme austerity policies pushed forward through a series of Memoranda of Understanding, signed between successive Greek governments and the lenders, the so-called troika (ECB, EU, IMF). These memoranda, the last one of which expired on the 21 August 2018, as it is obvious in retrospect, have had significant effects on the livelihoods of Greek people, including residents of Athens who form the focus of my chapter.

Over the months preceding this date, political debate within Greece and at EU level evolved around the post-memoranda era, with the Greek government arguing for a 'clear exit'. European Institutions, on the other hand, have started acknowledging 'the sacrifices of the Greek people' but left no room for an effective reorientation of policy or room for manoeuvre in the foreseeable future. Negotiations focused on the level of primary surpluses, on the non-performing loans (which the banks have to dispose of), and on pension cuts from the beginning of 2019, while proposals of the IMF for reduction of the unsustainable debt were fiercely opposed by Germany (Husson, 2018; Hadjimichalis, 2018). Thus, hopes for more flexibility and some leverage to plan for socially sensitive policies disappeared in the clauses of an exceptionally strict surveillance plan now in operation, continuing on the ideological fix on austerity which protracts recession and high unemployment rates.

This chapter discusses the unequal effects of crisis and austerity policies on Athens, based on past and ongoing research which includes bibliographical and statistical evidence, interviews with women from different social and ethnic backgrounds, participant observation in several urban neighbourhoods of the metropolitan area, newspaper clippings from three daily newspapers with extensive economic pages and electronic sites (for earlier presentations in English, see Vaiou, 2014, 2016). In the first section I summarise some aspects of the crisis in Athens; then, starting from the premise that behind figures and statistics lie embodied subjects, in the following section I discuss three major areas of argument between the creditors and the Greek government: property ownership and taxation, reduction of pensions, and continued austerity, from the perspective of particular people's experiences. Finally, I outline some ideas about bridging scales as a way into a better understanding of Athens in times of crisis.

Austerity and crisis in place

> The years 2010 to 2018 will go down in Greek history as an epic period of colonisation; of asset stripping and privatisation; of unfunded health and education; of bankruptcies, foreclosures, homelessness, and impoverishment; of unemployment, emigration, and suicide. These were the years of the three memoranda, or 'financial assistance programmes' accompanied by 'structural reforms', enacted supposedly to promote Greek 'recovery' from the slump and credit crunch of 2010. They were, in fact, a fraud perpetrated on Greece and Europe, a jumble of bad policies based on crude morality tales that catered to right-wing policies to cover up unpayable debts.
>
> (Galbraith, 2018, p. 1)

This quotation from a recent article by J. K. Galbraith (2018) summarises some of the effects of austerity policies which have been analysed by many researchers from different perspectives and political inclinations (Athanasiou, 2012; Karamessini, 2013; Skordili, 2013; Vradis & Dalakoglou 2011). Meanwhile, the crisis in/of Greece has made front-page news in EU media as the country became a site for experimentation on a number of frontal attacks on whatever constituted social citizenship and public provision. 'Salvation' of Greece, which was practically salvation of the German and French banks exposed to risky investments in the country,[1] has been used by the creditors as a legitimation of a whole campaign vilifying the undeserving Greeks as lazy, living beyond their means, etc. (Hadjimichalis, 2018). Against all evidence, little attention was paid to the fact that Greece and the undeserving Greeks were hard hit by impoverishment, high unemployment, loss of assets, and dramatic deterioration of living conditions, destruction of a productive structure based largely on small firms and self-employment, as well as the rise of the far-Right. Apart from academic and media analysis, Eurostat data and yearly reports by the Research Institute of the Confederation of Trade Unions (INE/GSEE), the Research Institute of the Confederation of Employers and the Bank of Greece provide ample evidence, to which I refer below.

The outcomes of 'salvation' are more than visible in Athens: closed shops and street-level uses on very central streets and in every neighbourhood are perhaps the most visible aspect of the crisis, testifying to the 75,600 firms of various sizes which have closed down (INEMY, 2017). New openings are quite temporary and short-lived (cafés, fast food ventures, nail salons, hairdressing, and so forth). However, unused built volume is much more significant and includes entire buildings and thousands of flats and offices dispersed on the upper floors of apartment buildings (Balabanidis, Patatouka, & Siatista, 2013). Homeless people in the arcades of unused buildings, drug addicts, sex workers (often victims of trafficking), beggars in many parts of the metropolitan area are an everyday part of the urban scene of the crisis-ridden city. The social violence implied in these developments is further aggravated by the aggressions of the far-Right and its claims to territoriality in particular neighbourhoods, mainly those with a high proportion of the migrant population (for a more detailed discussion see Vaiou, 2016; Kandylis, 2013).

Less visible, yet by now well-known aspects of the crisis include high unemployment and impoverishment, deterioration of living conditions (measured on the index of material deprivation),[2] care gaps, loss of assets of ever-larger sections of the urban residents, in a context in which GDP has fallen by 23.6 per cent (to 1964 levels), the volume of production has fallen by 23.5 per cent, incomes have dropped (Pantazis & Psycharis, 2016), while social and spatial inequalities have amplified and labour rights and democratic processes have been curtailed.

According to a recent survey,[3] in Attica, the region of Athens, households have lost on average 26 per cent of their income and 37.5 per cent of their assets since 2010. In this survey, 42 per cent of households declare that they cannot afford to heat their homes (see also Vatavali & Hatzikonstantinou, 2018), 13 per cent cannot provide adequate meals on a regular basis (including meat, fish, or chicken every other day), 52 per cent could not face unexpected expenses of 500€, while 60 per cent of those unemployed are unemployed for more than two years, and only 10 per cent of them receive unemployment benefit of 360€ per month; delays in payments of interests and rents and impossibility to pay utility bills are on the increase. In 2016, 300,000 households and 1,053 schools with 152,000 pupils were entitled to food support.

With the spread or generalisation of flexible and precarious forms of employment, including unregistered ones, the proportion of people at risk of poverty and exclusion rose to 35.6 per cent in 2016. The figure was then reduced technically, by lowering the poverty line to 4,500€ for one-person households per year (7,170€ in 2010), thereby reducing the number of those falling under the poverty line or at risk of that. Impoverishment is linked to many parameters, most prominently (long-term) unemployment and reduction of real/disposable incomes. After a peak in 2013, unemployment in 2017 was still at 21.5 per cent and has dropped to 18.5 per cent by the end of 2018. However, if discouraged unemployed, non-voluntary part-timers, and occasionally employed people are taken into account, the figure is estimated to reach 29.6 per cent (INE/GSEE, 2018). At the same time, income inequalities have become sharper both socially and spatially (Pantazis & Psycharis, 2016) while direct taxation remains disproportionately high and unequal (the lowest 20 per cent pays more than the highest 20 per cent), thereby increasing inequality between 2010 and 2015.

From 'the Greek crisis' to everyday women and men in the city

Statistical evidence and quantitative surveys are important in order to assess the overall extent and the dramatic effects of the crisis and austerity policies across the country in a language that is widely common and acceptable. Together with their connections with broader processes such evidence shapes what I would call 'the big picture'. My interest, however, is to move beyond general aspects of 'the Greek crisis' towards a more grounded engagement with the ways in which everyday women and men develop individual and collective practices of coping with, and sometimes resisting, austerity. Such an engagement involves a different theoretical perspective, informed by the diversity of the urban experience, as well as crossing across geographical scales and time-frames. It also involves a methodological approach which values as an equally relevant research input those voices that are usually silenced or marginalised when urban matters are debated. In this context, I summarise here material from the life-stories of women in Athens who have recently experienced deep changes in their everyday lives as a result of austerity policies.[4] Using the stories of embodied subjects does not methodologically imply some claim of authenticity (Crang, 2002), nor is it an idiosyncratic particularity. It is an important scale that cannot be ignored when we deal with understandings of 'the' crisis.

Pension cuts and crises of care

As I have already mentioned, one of the main issues debated in view of the post-memoranda era is the level of primary surpluses, which represent the capacity of the Greek economy to pay back its creditors. Set at 3.5 per cent, these surpluses mean that structural reforms will continue to be implemented (until 2060). A significant component of these reforms are more pension cuts, in order to ensure the viability of the social security system, a reform that the Greek government argued against and managed to modify at the end of 2018 – yet another unilateral act which caused a lot of argument during the first preliminary assessment by the surveillance team. Pensions, however, continue to be a main source of monetary income for households with more than one unemployed members and had already been cut several times since 2010. At the same time, pensions sustain arrangements for elderly care in an ageing society with 14.8 per cent over 70, as my first story indicates.

Christiana (interviewed in 2011 and 2016), was 56 years old at the time of the first interview, a public servant, divorced, and struggling to keep her standard of living (also discussed in Vaiou, 2016). She used to employ Athena, a migrant woman from the Ukraine, as a live-in carer to look after her father who is a widower and lives on his own in the same apartment building. In 2004, employing a live-in carer seemed like an affordable alternative that did not conflict with Christiana's ideas of appropriate care, as opposed to a home for the elderly (see Bettio, Simonazzi, & Villa, 2006; Mingione, 2009). This arrangement was common among a wide range of households with stable income and made room for Christiana to manage her time without feeling continuously preoccupied and guilty. The arrangement relied on the one hand on the abundant availability of low-paid migrant women, one of whom is Athena, and on the other on old-age pensions, like the one of Christiana's father. Those pensions were low but ensured the material conditions of this type of personalised care, with little contribution on the part of the family, in this case Christiana and her brother.

This model of care has gradually become unviable with the recurrent cuts of pensions and salaries since 2010 and the restructuring of the National Health System (another victim of structural reforms), the latter meaning that pensioners pay higher contributions for medical exams and for medicines 'of which the old man needs a lot', as Christiana says.

> My father's pension is not enough anymore to pay Athena and support a separate household. My brother is on the verge of being made unemployed – can you imagine, at the age of 58? (…) And his wife had to close her little corner shop – she was not earning any money (…). Thank god their children had both gone to study in England and stayed there. Anyway, he cannot help. And my salary has gone down tremendously. I do not get any more overtime or extra salaries – that is what I was giving for my father.

Christiana's daughter Athena, a 26-year-old graduate from the department of sociology and unemployed at the time of the second interview in 2016, comes to the assistance of her mother and proposes to look after her grandfather, while her mother is at work.

> But what about after work or at night? And what about Athena? I had to dismiss her after eight years – and I still worry about her. I have introduced her to other people, but everybody fears cuts and taxes – they try to cope on their own.

For Athena, losing this job means more than just becoming unemployed; it also means that she is at risk of losing her work permit, the renewal of which depends on the payment of social security contributions, and then being unable to send money back to her family in the Ukraine – which poses questions about seldom mentioned interconnections and hidden effects of the Greek crisis at different scales and across borders.

Owning a house in a middle-class suburb

In Athens what has been called a 'Southern' model of housing prevails (Allen et al., 2004), which includes high levels of homeownership (over 65 per cent in Athens) and a very small percentage of rentals, subsidised or not, a large sector of second homes, strong involvement of the family in younger members' access to homeownership, which until 2007 has kept the number of mortgages low (less than 20 per cent of homes; Emmanuel, 2015). What is important to underline here is that these features cut across class lines or income inequalities and, after

2000, also involve migrants of the 1990s who demonstrate low but increasing property ownership percentages (Balabanidis, 2016). The story of Despoina is inscribed in this context.

Despoina and her husband (interviewed in 2006 and 2017) are both around 60, they have two children (aged 23 and 25) and were both unemployed for more than three years in 2017. They used to run an evening school ('*frontistirio*' in Greek) for languages near their home. For this they had taken a small business loan which is now 'red', i.e. non-performing. With salary cuts and generally reduced incomes, demand for their services went down and, after a while, the middle-aged couple was obliged to close their business, like many other small businesses in the neighbourhood. But in 2017 they were too young to qualify for a pension. In common with many people of her class and age, Despoina had inherited from her parents the house they live in, a single family house in a suburb near the port of Piraeus. She says:

> I have grown up believing that even if worse comes to worse I will still have a roof over my head. But now (…) I feel we cannot even count on that. And we are trapped in it. We cannot pay the taxes, we cannot pay the loan we have not paid for three years now. We had to choose, either we do not eat or we do not pay the taxes (…). I do not dare answer the phone if I don't know who is calling, in case it is the bank and they tell me that the house is no longer mine (…). They will grab it and I will still be in debt you know.

With the increase of taxation on property, a major requirement of the creditors, Despoina and her husband, by now long-term unemployed and not entitled any longer to unemployment benefits, with only occasional earnings from small jobs, they cannot pay their taxes or loan instalments. They cannot even move to a smaller and less costly house and rent this property to live on, because that would require major repairs they cannot afford, plus the cost of moving. They spend winters without heating; they struggle every month to make ends meet and pay at least part of the electricity and water bills (after making arrangements with the respective companies). In addition, they are trapped, as she says, in an area where there are no low-cost supermarkets, no adequate public transport, no job opportunities in the vicinity.

> When my mother was still alive, her pension helped us out, at least with food. Now we have to economise even on that. I never expected I would have to go late to the open market and collect the food that greengrocers would throw away; or to ask for credit from the local grocery (…). And we cannot support at all our children who study in England. They have to work and study – they have difficult times as well.

Their only hope for some improvement is that they will reach retirement age and will be eligible for the national pension (€384), if they manage to pay a few more years into the pension fund.

Precarious lives in persisting austerity

Already before the end of August 2018, it became clear that the possibilities of the Greek government for reorientation of policy towards development priorities and social goals would be minimal. Far from leaving room to support the highly diverse and impoverished population of the metropolitan area, the creditors made it clear that little would change in terms of priorities, even after the memoranda expired. Continued austerity, as it is by now well-documented (Hadjimichalis, 2018; Coulmin Koutsaftis, 2018; INE/GSEE, 2018), hits hard the least powerful, among which many migrants of the 1990s, who constitute 10 per cent of the population of Athens and more than 25 per cent in some central neighbourhoods. In this context, far-Right discourse and racism find fertile ground to grow while material aggressions are only temporarily suspended due to the ongoing trial of the leaders of the neo-Nazi party Golden Dawn. Together with the EU controversial attitudes on refugees, practices of living together with 'others' in the same buildings and urban neighbourhoods are in danger. Even people living in Athens for more than 30 years or born and grown there feel unwelcome. The story of Lidia unfolds in this context.

Lidia is in her early 30s (interviewed in 2006 and 2017), she came from Fieri (Albania) to Athens in 1993 at the age of nine, together with her parents. She attended the Greek public school in a neighbourhood in central Athens, fast overcoming language difficulties, thanks to the assistance of her teachers. Her mother (a civil servant in Fieri who works as a live-out cleaner in Athens) urged her to go for higher education and Lidia took a degree in regional planning from the University of Thessaly. In order to finance her studies, she got some little money from her family and, mainly, worked in cafés and bars in the city of Volos.

> In any case, I had to work after the age of 18, in order to collect the necessary social security stamps and be able to renew my residence permit. Only recently has this thing changed and I could take long-term resident status and avoid this killing bureaucracy.

Upon graduation, she was very happy to find a job in a private planning consultants' office in Athens (2006–2013), she got married and had a daughter who was four at the time of the 2017 interview.

> My daughter was born at the time I was made unemployed. Actually, we decided to have a child while I was looking for another job – and could not find one, as you know. Small planning offices were closing at that time, as the crisis persisted and there were no new contracts around. Then started the problems (…). You know, I don't mind doing any kind of job – my mom has taught me that work is not a shame (…). But with small jobs and occasional day-earnings, the money is not enough.

When her husband lost his job as well, in 2016, she felt very insecure and, after difficult discussions, the young family moved to her parents' house, in conditions of relative overcrowding.

> I had to decide what was more important, to have food on the table or to keep our 'independence'– and my constant fear that we will not be able to pay the rent and be evicted (...). Sokol (her husband) felt humiliated; he could not somehow cope with the idea that he could not support his family (...). The situation became bad between us. We divorced about a year ago.

Lidia now works in a restaurant kitchen only weekends, goes for food assistance to the municipality, and to the neighbourhood solidarity initiative, she collects vegetables for free when the weekly open market in the neighbourhood is about to close and the greengrocers give away what they have not sold. She makes every effort not to depend entirely on her parents who, in any case, are in a difficult state as well. In the new extended family, only Lidia's mother has a regular job; her father works occasionally in construction and hopes that the sector will pick up in the near future.

Lidia lives in emotional strain and physical exhaustion, linked to little income, job insecurity, and all the efforts to collect food, to identify services she could be eligible for free (e.g. health or childcare), as well as a divorce following the decision to move back to her parents' house. She feels socially devalued and humiliated, in constant fear and 'trapped' in Athens and in 'her neighbourhood', but also fear of things getting worse for 'others' like her.

> Where am I to go? Back to Fieri? We don't have family there anymore. Besides, all those who somehow return, they come back here, you know. Things are not easy there either (...) and where do I take Mosa (the daughter)? She is not a package – she only knows this place, this neighbourhood, this language.

The dire conditions in which Lidia lives the crisis in Athens are common to many people who had migrated to the city after 1989 and have lived here for almost 30 years by now. For them, 'here' is home, networks, every day and longer-term life plans; it is the place where their parents have struggled to 'make life good', as one woman has told me. 'There' (where their parents came from) is not part of their experience and going there is not an option; it would mean a new migration project and the struggle to make yet another start. For them the repercussions of austerity policies hit harder key aspects of everyday survival and longer-term well-being as the safety net of the family can provide only marginal resources.

Crossing scales or what do these stories tell us?

After more than eight years of extreme austerity and sometimes punitive impositions on the 'undeserving Greeks', few believe that change is possible in the foreseeable future. Pessimism about possible prospects is further reinforced by a 'newspeak', like the one described in George Orwell's (1949) famous novel. This 'newspeak', introduced by several high standing officials who recently also make references to the 'sacrifices of the Greek people', remains on a very general scale of reference, obscures social effects, and rationalises the failures of imposed priorities and policies which have practically destroyed Greece. The following three examples, drawn from Compliance Reports produced by the European Commission during 2018, are a case in point (see Husson, 2018).

'Rationalising overall expenses' is a reference to the further dismantling of social welfare, which the current government tries hard to reinstate in the cracks and margins of surveillance and is continuously accused of unilateral actions. Here, pension cuts seem an emblematic banner against all evidence that further reduction of incomes will not only send households deeper under the poverty line and create care deficits, but it will also send consumption down and jeopardise possibilities for recovery.

'The fight against a culture of non-payment' (of taxes and loan instalments) refers to banks that are free to confiscate people's assets for any liability over €1,000. Owning a house through inheritance, family transfer, savings, and more recently also loans (and combinations of the above) is considered beyond the means of 'lazy Greeks', and probably even more lazy Athenians, who have aspired to homeownership as a means of security at times of crisis.

'Evolution towards a more modern economy' covers the demand for high primary surpluses and continued austerity policies that destroy the productive structure of a city, which indeed is not conforming to northern European standards. Such policies also produce high unemployment, lead highly qualified youth to emigration, nourish far-Right discourse and aggressive practices, and jeopardise social cohesion and living together with a wealth of 'others' in urban neighbourhoods.

The above examples characterise the way in which 'the Greek crisis' has been represented in a public discourse which has only very recently started to suggest that perhaps the measures imposed on Greece have been too strict, based on wrong indicators and recurrently failed forecasts. The post-memoranda praises of officials like Donald Tusk, Klaus Regling, Mario Senteno, Jean Claude Juncker, and others (through tweets on 20 August 2018) have little to say about

> those who lost their homes and ended in the streets, about the thousands who committed suicide cracked by the pressure of the new situation (…), about all those, the silent ones, who isolated themselves socially unable to bear the loss of their dignity.
> (Keep Talking Greece, 2018)

The passage from general data and theoretical conceptions of 'the Greek crisis' to concrete places and to the experiences of particular embodied subjects (and back) is not an easy project. It requires crossings of geographical scales, each of which reveals different and, I would argue, equally important facets of the issues under study. But such crossings help carry the argument forward in two directions. First, they help understand the multiple determinations of an otherwise unqualified and almost generic conception of crisis which leaves no escape route. Second, they help shape an approach which consciously oscillates between levels of reference that are usually kept apart: on the one hand, discourses and explanations constituted by 'big pictures' and global analyses and, on the other hand, urban space and the spatialities produced through the bodily presence and everyday practices of individuals and groups (see Simonsen, 2008).

The women whose stories I have quoted, like many others, have seen their personal life projects overturned by the crisis albeit in ways which differ in significant ways. They have to survive with lower or no income, curtailed services, increased depression, and insecurity as well as a heavier burden of everyday care work. Their lives now include all sorts of improvisations and detailed adaptations necessary for everyday survival. As the stories of these ordinary women also tell us, living with the crisis in the city and with a multiplicity of possible trajectories and life choices – constituting a progressive sense of place as Doreen Massey (1994, 2005) has been urging us – is more than a theoretical conception. It is a major stake, a process of familiarisation with difference/s and otherness, which includes controversies, requires investment of time and labour, both material and emotional, abundantly contributed by bodies which usually do not matter in dominant discourse. Mobile and immobilised, legitimate and illegitimated, differentiated, gendered, sexualised, racialised, 'othered' bodies constitute resources necessary for coping with, but also protesting and resisting, austerity and crisis through inconspicuous everyday practices.

A view from this scale, linked in multiple ways to many other scales (local, national, European, international), reveals areas of knowledge that would otherwise remain in the dark, as feminist geographers have forcefully argued for many years. Stories which connect global processes with concrete bodies enrich our understandings with more complex and more flexible variables and inform the 'big pictures' – and not only the reverse. Such a theoretical and methodological approach is important not only theoretically but also politically at the present conjuncture, because it provides a vantage point from which to re-examine the meanings and practices of 'doing politics' and re-evaluate claims of access, visibility and participation in urban life and pathways out of the crisis.

Notes

1 The gigantic loans directed to Greece in 2010–2012 were mainly destined to payoff German and French banks, by then exposed to risky investments in public and private debt and threatening the stability of the euro. The 'salvation of Greece' was then turned into a series of bilateral loans between the country and each EU member (including poorer ones) and followed by a series of burdensome bailout packages whose stated goal was to reinstate the competitiveness of the economy (Coulmin Koutsaftis, 2018).

2 The index of material deprivation, following elaborations of the SILC surveys (Statistics on Income and Living Conditions), measures the (enforced) inability to pay unexpected expenses, afford a one-week annual holiday away from home, a meal involving meat, chicken, or fish every second day, adequate heating of a dwelling, durable goods like a washing machine, colour television, telephone, car; it also includes being confronted with payment arrears for mortgage, rent, utility bills, or other loan payments. See https:// ec.europa.eu/statistics-explained/index.php/Glossary:Material_Deprivation. In Greece, these features concern not only the poor, but also part of the non-poor population.

3 The survey was commissioned by the Regional Government of Attica in cooperation with the Labour Centre of Athens (EKA) and conducted between 22 November 2017 and 15 December 2017, by the private company Marc, with a sample of 1,501 people, 1,001 of whom were employed and 500 unemployed (see Marc, 2018).

4 The material in this section comes from the following research projects and from 'revisiting' participants in those projects: Dina Vaiou (coordinator) 2005–2007 *Intersecting patterns of everyday life and socio-spatial change in the city. Migrant and local women in the neighbourhoods of Athens* (Ministry of Education, Pythagoras II Programme); Maria Stratigaki (coordinator) 2007–2011, *Gender, Migration and Intercultural Interactions in the Mediterranean and South East Europe: an interdisciplinary approach (GeMIC)* (Panteion University, Athens – member of the Greek team on gender and urban space) (European Commission – 7th Framework Programme); Maria Stratigaki (coordinator) 2009–2012 *Transnational Digital Networks, Migration and Gender (mig@net)* (Panteion University, Athens – scientific adviser to the research team) (European Commission – 7th Framework Programme); Dina Vaiou (coordinator) 2013–2018 *Gendered aspects of the crisis in Athens – Revisiting research places and subjects* (NTUA).

References

Allen, J., Barlow, J., Leal, J., Maloutas, T., & Padovani, L. (2004). *Housing and welfare in southern Europe*. Oxford: Blackwell.

Athanasiou, A. (2012). *The crisis as a 'state of exception'. Critiques and resistance*. Athens: Savvalas (in Greek).

Balabanidis, D. (2016). *Geographies of migrant settlement in Athens, 2000–2010. Housing trajectories and access to home-ownership*. Unpublished PhD Dissertation, Department of Geography, Harokopio University Athens (in Greek).

Balabanidis, D., Patatouka, E., & Siatitsa, D. (2013). The right to housing during the crisis in Greece. *Geographies*, *22*, 31–42 (in Greek).

Bettio, F., Simonazzi, A., & Villa, P. (2006). Change in care regimes and female migration: the 'care drain' in the Mediterranean. *Journal of European Social Policy*, *16*(3), 271–285.

Coulmin Koutsaftis, M. L. (2018). Les emprunts non performants, la situation du parc immobilier en Grèce et les saisies des résidences principales. Retrieved from www.defenddemocracy.press/appauvris-par-les-memoranda-les-grecs-vont-perdre-tous-leurs-biens.

Crang, M. (2002). Qualitative methods: The new orthodoxy. *Progress in Human Geography*, *26*(5), 647–655.

Emmanuel, D. (2015). Social aspects of access to home ownership. In T. Maloutas & S. Spyrellis (Eds.), *Athens social atlas*. Digital compendium of texts and visual material. Retrieved from www.athenssocialatlas.gr/en/article/access-to-home-ownership.

Euro Memo Group (European Economists for an Alternative Economic Policy in Europe). (2018). Prospects for a popular political economy in Europe. Retrieved from www.euromemo.eu.

Galbraith, J. K. (2018). The Greece bailout's legacy of immiseration. *The Atlantic*. Retrieved from www.theatlantic.com/international/archive/2018/08/greece-bailout-imf-europe/567892.

Hadjimichalis, C. (2018). *Crisis spaces. Structures, struggles and solidarity in Southern Europe*. London and New York, NY: Routledge.

Husson, M. (2018). Un long calvaire s'annonce pour la Grèce. *Alternatives Economiques*. Retrieved from www.alternatives-economiques.fr/un-long-calvaire-sannonce-grece/00085215.

INE/GSEE (Institute of Labour of the General Confederation of Trade Unions) (2018). *Annual report 2017: The Greek economy and employment*. Retrieved from www.inegsee.gr (in Greek).

INEMY (Institute for Commerce and Services) (2017). Survey on closures of enterprises on central commercial streets, Sept. 2017. Retrieved from www.inemy.gr/ekdoseis/erevnes.

Kandylis, G. (2013). The space and time of migrant resentment in the centre of Athens. In T. Maloutas, G. Kandylis, M. Petrou & N. Souliotis (Eds.), *The centre of Athens as a political stake* (pp. 257–279). Athens: EKKE and Harokopio University (in Greek).

Karamessini, M. (2013). Structural crisis and adjustment in Greece. Social regression and the challenge for gender equality. In M. Karamessini & J. Rubery (Eds.), *Women and austerity: The economic crisis and the future for gender equality* (pp. 165–185). London: Routledge.

Keep Talking Greece (2018). Lost in hypocrisy: Greece's lenders celebrate end of bailout programs. Editor's note. Retrieved from www.keeptalkinggreece.com/2018/08/20/greece-end-bailout-program-greek-opinion/.

Marc (Marketing Research and Communication) (2018). Living in Athens. Summary of findings. Retrieved from www.keeptalkinggreece.com/2018/02/23/athens-attica-survey-living-condtions/.

Massey, D. (1994). *Space, place and gender*. Cambridge: Polity Press.

Massey, D. (2005). *For space*. London: Sage.

Mingione, E. (2009). Family, welfare and districts. The local impact of new migrants in Italy. *European Urban and Regional Studies*, *16*(3), 225–236.

Orwell, G. (1949). *1984*. London: Secker & Wardburg.

Pantazis, P., & Psycharis, G. (2016). Residential segregation based on taxable income in the metropolitan area of Athens. In T. Maloutas & S. Spyrellis (Eds.), *Athens social atlas*. Digital compendium of texts and visual material. Retrieved from www.athenssocialatlas.gr/en/article/income-groups/.

Simonsen, K. (2008). Practice, narrative and the 'multicultural city': A Copenhagen case. *European Urban and Regional Studies*, *15*(2), 145–158.

Skordili, S. (2013). Economic crisis as a catalyst for food planning in Athens. *International Planning Studies*, *18*(1), 129–141.

Vaiou, D. (2014). Is the crisis in Athens (also) gendered? Facets of access and (in)visibility in everyday public spaces. *City*, *18*(4–5), 533–537.

Vaiou, D. (2016). Tracing aspects of the Greek crisis in Athens: Putting women in the picture. *European Urban and Regional Studies*, *23*(3), 220–230.

Vatavali, F., & Hatzikonstantinou, E. (2018). *Geographies of energy poverty in Athens in crisis*. Athens: angelus novus (in Greek).

Vradis, A., & Dalakoglou, D. (2011). *Revolt and crisis in Greece*. Oakland/Edinburgh: AK Press.

8 The *right to the city* after Grenfell

Gareth Millington

Introduction

The right to the city has been one of the great rallying cries of urban protestors and activists across the globe over the last 50 years yet it has hardly been evoked in the aftermath of the Grenfell fire in London in June 2017 which killed 72 people. There are many possible reasons for this. On a surface level, there appears to be a discrepancy between the generality of the right to 'the city' and the specificity of the fire itself; a further incongruity exists between the grieving and political organising in process among local communities in North Kensington and the relevance of high-brow theory to these processes (Shilliam, 2019). In addition, the universalistic appeal of the right to the city fails to illuminate the racialised and gendered experiences of many who lived in Grenfell and its surroundings (for a set of such accounts see Bulley, Edkins, & Al-Enany, 2019). A further problem, addressed here, is that the Lefebvrian formulation of the right to the city says nothing about protecting the safety of urban inhabitants, choosing instead to emphasise themes of freedom, playfulness, and creative expression. If the right to the city is so incongruent to an understanding of Grenfell and its attendant struggles, then serious questions about its contemporary political value need to be asked. To that end, this chapter outlines some of the ways that the right to the city – as theory and praxis – may be reinterpreted, renewed, or reimagined post-Grenfell in such a way that better supports 'those who think and work and survive on front lines' (Shilliam, 2019, p. 197).

In its distrust of the state, bureaucracy, the police, and urban planners Henri Lefebvre's notion of the right to the city is imbued with the exuberant spirit and mythology of Paris 1968. In fact, Bodnar (2013, p. 76) argues that May 1968 is an event whose 'script was written partly by Lefebvre'. Lefebvre's relationship with 1968 was actually more fraught; for example, he was dissatisfied with the protestors' vague demands, while they were frustrated with Lefebvre for not putting his ideas – taught with such passion at Nanterre University where the protests started – into practice (Merrifield, 2002, p. 86). The right to the city also resonates with another moment in Parisian history, the Commune of 1871, which Lefebvre (2003, p. 188) refers to approvingly as 'a revolutionary festival and festival of the Revolution, a total festival'. Understanding the right to the city as a festival is central to Lefebvre's writing on the subject. The right to the city involves a joyful refusal and subversion of the 'bureaucratic society of controlled consumption' (Lefebvre, 2000). Certainly, Lefebvre is dismayed by planning rationalities and industrial production (Lefebvre, 1996, p. 149); he argues that capitalism and modern statism work together to crush the creative capacity of the urban *oeuvre* (Kofman & Lebas, 1996, p. 2). In contrast, the right to the city reaches out and steers 'towards a new humanism, a new praxis, another man [*sic*], somebody of urban society' (Lefebvre 1996, p. 150). It is a 'renewed right to urban life' based around an understanding of the urban as a place of encounter and exchange, and the prioritising of use value over exchange value (Lefebvre 1996, p. 158). Lefebvre was unambiguous in asserting that the right to the city has particular bearing and significance for the working-class, for those who are dispossessed from the centre of the city and expropriated from the co-produced social and cultural benefits of urban life (Lefebvre 1996, p. 179). Lefebvre writes that '[only] this class, as a class, can decisively contribute to the reconstruction of centrality destroyed by segregation and found again in the menacing form of *centres of decision-making*' (Lefebvre 1996, p. 154, emphasis added). Decision-making is, of course, a euphemism for the power of state bureaucracy, whose rationality, argues Lefebvre (Lefebvre 1996, p. 127), sits 'above' the city, viewing the city as an instrument or a means, while refusing it as an *oeuvre* in its own right.

The Grenfell Tower fire prompts another evaluation of the right to the city, especially in relation to Lefebvre's disdain for bureaucracy and technocracy. In short, the problem is that Lefebvre's conception of this right does not address the amplified demand – in neo-liberal and austere times characterised by what Tombs (2017) calls the 'undoing of social protection' – to live in safe housing. Indeed, the Grenfell fire highlights the role that the state *is still expected by citizens to play* even in the most modest lay understandings of the right to the city. It is difficult to conceive of a right to protection without a role for the state, or at least a functioning, radical bureaucracy. There is comparability here with Madden and Marcuse's endorsement of a radical right to housing: 'The state is clearly part of the problem, and yet is absolutely necessary for any solution' (Madden & Marcuse, 2016, p. 198).

Another problem is that in neo-liberal (and authoritarian) times, urban self-determination – which Lefebvre offers as the only viable alternative to the state – unwittingly offers an alibi for the continued withdrawal of the state from certain aspects of bureaucratic and social support, or as a way of justifying the social harms that this withdrawal causes (i.e. it restores responsibility to the individual). Bodnar suggests that Lefebvre's right to the city too easily 'suits [neo] liberal urban justice discourses of *participation*' (Bodnar, 2013, p. 81) added emphasis). In practice, in relation to urban safety, poorer urban residents tend either to be subject to harassment by an intolerant police force (Vitale, 2017) or their demands, often via complaints to public bodies, are ignored or passed off as troublemaking (Renwick, 2019). Rarely do they participate in a manner consistent with the rhetoric. Of course, as a Marxist – albeit an unconventional one – Lefebvre envisaged a wholly different scenario whereby self-determination – or *autogestion* – could only be realised alongside the *withering away* of the state. The point is that post-Grenfell, many wish the state would accept responsibility rather than becoming ever more distant. Thus, it is imperative to consider what is gained (and lost) by incorporating the demand for the right to protection into understandings of the right to the city. If the right to the city – probably the most radical and seductive idea of the twentieth century in terms of imagining the relation between cities and citizens – cannot be mobilised in the aftermath of a catastrophe such as Grenfell, then it must be asked what confidence we should continue to invest in it.

The aim of this chapter is to consider the questions posed by the Grenfell fire in relation to both the right to the city and the 'political economy of urban safety' (Atkinson & Millington, 2018, p. 265). It is necessary to grapple with the idea and practice of a claim to protection – a claim articulated and made by Grenfell residents, their friends, and families before and after the fire – and to consider, by extension, the difficult question of *just how safe do we want or expect our cities to be*? Moreover, is it not more instructive to prioritise the *inequalities of exposure to harm* that characterise cities? As Madden puts it,

> [there] are aspects of urban environments and everyday life that can kill, either swiftly through catastrophic failure or ecological disaster, or slowly through illness or poor health. But the chances of being subjected to these conditions are distributed unevenly.
>
> (Madden, 2017)

The challenge then is to think through how protection may be incorporated into the right to the city but without losing the spirit, ethos, and utopian vision of Lefebvre's original formulation.

The chapter begins with an account of the Grenfell fire and the various reports that were written in its aftermath. This is followed by a return to the notion of the right to the city, focusing also on Lefebvre's preference for *autogestion*, or self-determination, over the authority of the state. As a counterweight to this argument, bureaucracy is then considered from a more positive angle, pointing not to the failings of bureaucracy that are now well publicised in the aftermath of the fire, but to how regulations, tests, controls, and enforcement – *properly exercised* – could have prevented the fire or protected citizens from its worst effects. The chapter then questions whether the policy notion of 'community safety' – originally formulated to counter the ravages of a decade of Thatcherism on the social fabric of cities – could, under a more radical and less managerial agenda, be incorporated into the practical and theoretical rubric of the right to the city. Finally, the chapter considers the relationship between politics and aesthetics in a reimagined demand for the right to the city.

The Grenfell Tower fire

> The dominant narrative of London is that of a *world class city*: one that for several years has topped PriceWaterhouseCoopers' global rankings for business opportunities, education, and quality of life. Less prominently featured in the consultancy firm's reports is the fact that no city in the global north internalises such a gulf between rich and poor (…).
>
> (MacLeod, 2018, p. 477)

Grenfell Tower was a 24-storey residential building comprised mainly of public housing units. It is situated in the Royal Borough of Kensington and Chelsea (RBKC), one of the wealthiest but most socially unequal boroughs in London. The tower was designed in 1967, approved in 1970 and completed in 1974, as part of the Lancaster West Estate. On 14 June 2017, the worst fire in UK peacetime since the nineteenth century, killed 72 residents of Grenfell Tower, most of whom were low- or modest-income Londoners and many of whom were recent or first-generation migrants to the city. Hundreds more were left injured, distressed, and/or homeless.

Up until 2017 the story of Grenfell is not untypical of a local authority tower block in London (see Boughton, 2018). In the 1980s and 1990s, due to a lack of funding and the political degrading of council housing and the symbolic denigration of local authority tenants, the Lancaster West Estate experienced deteriorating infrastructure, declining amenities, and deficient maintenance (Platt, 2017). In 1996, RBKC transferred responsibility for maintenance to the Kensington and Chelsea Tenant Management Organization (KCTMO). Worries about the loss of green space due to the construction of a new school and leisure centre close to the tower, a sense of estrangement from KCTMO's decision-making in addition to ongoing fears around safety caused residents, in 2010, to establish the Grenfell Action Group (GAG) (MacLeod, 2018, p. 468). Hodkinson (2019) describes a process of *de-municipalisation*, in play since 1979, whereby local authorities were prohibited from directly investing in and repairing their public housing stock unless this stock was sold or responsibility for its management and maintenance outsourced to commercial actors. Following a refurbishment of Grenfell that was managed by KCTMO and undertaken by Rydon Constructions in 2015–2016, residents raised concerns (to KCTMO) regarding risks caused by cost-cutting on materials, exposed gas pipes, the lack of a sprinkler system or integrated fire alarm, and the failure to provide more than one fire escape (Booth, Gentleman, & Khilali, 2017). Famously, a GAG blog post from November 2016, titled 'KCTMO-Playing with Fire!' forewarns the terrible events to follow:

> It is a truly terrifying thought but the Grenfell Action Group firmly believe that only a catastrophic event will expose ineptitude and incompetence of our landlord, the KCTMO, and bring an end to the dangerous living conditions and neglect of health and safety legislation that they inflict upon their tenants and leaseholders.
>
> (GAG, 2016)

As MacLeod (2018, p. 470) documents, residents from the Lancaster West and Grenfell community believed their concerns as social tenants were being ignored, that they were treated with contempt by authorities. This local erosion of democracy can be traced to central government,

> [not] least in that the original impulse to outsource erstwhile publicly managed services – at local and national levels – came from successive Thatcher-led governments in the 1980s: part of a wider endeavour to 'roll back' an ostensibly unwieldy and overly-bureaucratic state.
> (MacLeod, 2018, p. 474)

The feeling of being excluded from the local polity was exacerbated too by growing inequality in RBKC, with poorer residents sensing that they were victims of social cleansing and that even refurbishments were carried out for the benefits of the wealthy, who preferred to live without a sightline of 'ugly' social housing tower blocks like Grenfell (Boughton, 2019, p. 4).

If trust in the ability of the local state to provide adequate protection for residents of social housing was low before the fire, it was decimated afterwards. There were public protests and the resignations of Conservative RBKC leader Nicholas Paget-Brown and his deputy Rock Feilding-Mellen. Then Prime Minister Theresa May was heckled during a visit to Grenfell when to much astonishment she left without speaking to surviving residents or families of the missing. Later the Prime Minister would apologise for 'the failure of the state' (Kentish 2017), a statement not necessarily at odds with conservative favoured free-market policies of privatisation.

The Grenfell fire must also be understood in relation to austerity. As part of a political response to the financial crisis of 2007–2008, the Conservative–Liberal Democrat Coalition government drastically reduced the Communities Budget, with 50 per cent cuts to social housing and 40 per cent cuts to local government (O'Hara, 2015). Peck (2012, p. 651) argues, 'a financial crisis [was] transformed into a state crisis, and now that state crisis is being transformed into an urban crisis'. As MacLeod (2018, p. 475) explains, local authorities believed it was their job simply to spend as little as possible. This was evident in the cost-cutting efforts of RBKC and KCTMO during the Grenfell refurbishments (MacLeod, 2018, p. 473) and the firefighters who continued to risk their lives in a borough where fire cover had been cut by 50 per cent (Booth, Gentleman, & Khilali, 2017). For many on the Left, Grenfell was emblematic of the lethal consequences of government austerity. Rather than a 'tragedy', radical commentators such as Aditya Chakrabortty (and then Shadow Chancellor John McDonnell) prefer Engels' notion of 'social murder'. Chakrabortty writes:

> While in Victorian Manchester, Friedrich Engels struggled to name the crime visited on children whose limbs were mangled by factory machines, or whose parents were killed in unsafe homes. Murder and manslaughter were committed by individuals, but these atrocities were something else:

what he called social murder. (...). Over 170 years later, Britain remains a country that murders its poor. When four separate government ministers are warned that Grenfell and other high rises are a serious fire risk, then an inferno isn't unfortunate. It is inevitable. Those dozens of Grenfell residents didn't die: they were killed.

(Chakrabortty, 2017)

In this sense, the Grenfell fire was the predictable outcome of unregulated 'free-market' policies aimed at maximising capitalist profitability, a system that chooses to condemn many to live in *deadly* housing (Hodkinson, 2019, p. 5).

Materiality, testing, and the right to protection

The independent review of building regulations and fire safety (commissioned shortly after the Grenfell fire) states the regulatory system concerning high-rise buildings in England and Wales is not 'fit for purpose' (Hackitt, 2018, p. 11). Within the construction sector there has been a 'race to the bottom' regarding safety regulation leading to 'insufficient focus on delivering the best quality building possible, in order to ensure that residents are safe, and feel safe' (Hackitt, 2018, p. 5). The key issues underpinning system failure are ignorance of regulations and guidance on the part of those who are supposed to be experts, an indifference to quality caused by a motivation to build quickly and cheaply as possible (which also leads to residents' concerns and complaints being ignored), a lack of clarity on roles and responsibilities that has been exacerbated by fragmentation on the construction industry and inadequate regulatory oversight and enforcement tools: 'Where enforcement is necessary, it is often not pursued. Where it is pursued, the penalties are so small as to be an ineffective deterrent' (Hackitt, 2018, p. 5).

In another important report into the Grenfell fire (published in the industry journal *Inside Housing*) Apps, Barnes, and Barratt (2017) follow the paper trail that exposes the failure of building regulations. This takes us into the dry, detailed, technocratic realm that Lefebvre struggled to get excited about, yet its relevance for resident's safety soon becomes apparent. The failures listed in the report are largely attributed to the 'relaxed' attitudes of successive governments, post-1980, towards regulation of building materials and building control. The crucial legislation that led the way for these radical shifts was introduced in 1985 by Margaret Thatcher's government, where 306 pages of building regulations were swept away and replaced with just 24. The report also states that building regulations commissioned by government are open to interpretation when it comes to whether certain materials can be used as external cladding on buildings. Very few materials are ever banned outright; rather, a series of 'broad outcomes' in terms of safety must be reached for each building. It is up to the industry to interpret exactly how to meet such outcomes. Apps, Barnes, and Barratt (2017) also spell out how over time 'desk tests' of combinations of building materials have been preferred to actual tests of materials. BS (British Standards) 8414 fire tests are used to prove whether or not materials are suitable for high-rise buildings. One problem with such tests is that materials are tested individually and not in combination with the full range of products available on the market. Many combinations are therefore not tested at all although on the basis of individual tests they are cleared via a 'desktop study'. Neither the reports nor the methodology of safety tests are currently required to be made public. As Apps, Barnes, and Barratt state, combinations of materials have passed a desktop study only to subsequently fail a real-world test. Privatisation is also an issue. Key regulatory bodies such as the Building Research Establishment have been privatised. Moreover, building control has also been opened to competition which has created an incentive not to fail building schemes. Echoing the conclusions of the Hackitt Report (2018), one manufacturer of cladding is quoted as saying, 'It was a complete race to the bottom. They [building control] would approve everything to get [their] market share' (Apps, Barnes, & Barratt, 2017). The relaxation of building control was such that

> [in] 2016, shortly after the Grenfell refurbishment finished, the NHBC (National House Building Council) listed several combinations of cladding and insulation which it believed could be signed off without even the need for a desk top study. This included Celotex RS5000 insulation and 'Class 0' aluminium composite material cladding: the exact combination used on Grenfell.
>
> (Apps, Barnes, & Barratt, 2017)

Tests carried out after the fire have shown that the panels used in the Grenfell refurbishment failed with every combination of insulation it was tested with and '[in] the specific combination used on Grenfell, flames ripped through a nine-metre rig in less than 10 minutes' (Apps, Barnes, & Barratt, 2017). The mistaken belief, prior to Grenfell, was to rely upon the fire-resistant surface to the aluminium panel without considering the flammable materials, such as plastic polyethylene or polyisocyanurate insulation, which may be loaded between or behind the aluminium surface.

In his 'expert report' presented to the Grenfell Tower Inquiry in October 2018, Professor Luke Bisby from Edinburgh University also focuses his attention on the flammability of the *combination* of materials used in the aluminium cladding panels. The primary cause of fire spread was the PE (polyethylene) filler material found within the aluminium composite material (ACM) rainscreen cladding cassettes used on the exterior of Grenfell Tower (Bisby, 2018, p. 3). Bisby (2018, p. 75) explains, 'Aluminium Composite Materials (ACMs) are composite panelling products typically comprised of two layers of aluminium sheeting separated by a filler material'. The precise material used during refurbishments was Reynobond PE Composite Rainscreen system. Reynobond ACM is an aluminium panel consisting of two flat aluminium sheets, each 0.5mm thick, that are thermally bonded to both sides of a filler material.

According to Bisby, these panels are the 'critical material' in explaining the escalation of external fire spread. He writes:

> The available evidence strongly supports a hypothesis that the presence of aluminium composite panels with polyethylene filler material, over large areas of the external surface of Grenfell Tower, was the primary cause of upward vertical fire spread, downward vertical fire spread, and horizontal fire spread. The ACM product used on Grenfell Tower incorporates a highly combustible PE polymer filler which melts, drips, and flows at elevated temperature. The polyethylene filler material is expected to release large amounts of energy during combustion, to rapidly lose its mechanical properties, and to cause separation of the ACM panels and cassettes.
>
> (Bisby, 2018, p. 260)

Although he presents a scientific report, Bisby (2018, p. 3) stresses the importance of understanding *how* a polyethylene-filled rain-screen product could have been installed on a building of this occupancy and height in England. This is especially troubling he argues since,

> [these] hazards were well known and documented within the technical literature (…), within the available statutory design guidance (…), and via on line/popular media and press *prior to the design and construction of Grenfell Tower's refurbishment cladding system.*
>
> (Bisby, 2018, p. 261 emphasis added)

Bisby's report also discusses the cladding tests carried out following the Grenfell fire by the Department of Communities and Local Government (DCLG). A test of the combination of materials used on the Grenfell refurbishment 'was terminated 8 minutes 45 seconds after ignition due to rapid escalation of the fire, which spread upwards over and within the external cladding system' (Bisby, 2018, p. 172). Bisby continues:

> It should also be noted that to 'pass' a BS (British Standards) 8414–1 test (noting that 'pass/fail' terminology is not actually used to classify BS 8414 testing), a cladding system is required, amongst other things, to not experience rapid escalation of vertical fire spread within a period of 30 minutes – thus the [test] outcome must be considered as a particularly unsatisfactory result as regards escalation of external fire spread. Given this result, and the available technical literature, it is highly unlikely in my opinion that any realistic external cladding system which uses an ACM rainscreen with a non-fire-retarded PE filler could be expected to pass a BS 8414–1 test.
>
> (Bisby, 2018, p. 172)

The reports on the Grenfell fire published thus far point to the importance of (1) political-economic context, especially the decades-long assault on building regulations (and control), health and safety regulations, and other forms of 'red tape'; (2) an existing regulatory system on building materials that is ambiguous and open to interpretation, so as not to impede business practices and profit-maximisation in the construction industry; and (3) following privatisation, the declining significance and public scrutiny of real-world British Standards fire tests of combinations of building materials.

All this evidence must be understood in the context of what Tombs (2017) calls the 'undoing of social protection'. In his study of environmental health regulation in the north of England, he argues that neo-liberalism and austerity have caused a decline in inspections. Moreover, even when inspections are carried out there is little political support for authorities to pursue enforcement and prosecutions. It has reached the point where 'public health and protection is being eroded' (Tombs, 2017, p. 136). Tombs continues:

> Taken together, the trends set out above may mark the beginning of the end of the state's commitment to, and ability to deliver, social protection. What began as a neo-liberal policy turn to 'better regulation' then become turbo-charged under conditions of austerity, where the state claims that it cannot afford to enforce law (...). The subsequent institutionalisation of non-enforcement of law sends a green light to business that its routine, systematic, widespread social violence is to be tolerated, allowing private business to externalise the costs of its activities onto workers, consumers, communities, the environment. It further diminishes the quality and longevity of lives of those with the least choice about where they live, what they do for a living.
>
> (Tombs, 2017, p. 139)

As Cooper and Whyte state, 'attacks on the publicly funded services that are supposed to protect people in almost all spheres of social life have produced profoundly violent outcomes' (Cooper & Whyte, 2017, p. 3). Moreover, it is important to consider 'the assemblage of bureaucracies and institutions [and materials] through which austerity policies are made real' (Cooper & Whyte, 2017, p. 3). Cooper and Whyte explain the harm caused by austerity policies, including the erosion of health and safety regulations as part of the normalisation of the violence of the state. While this is true, it is not the state per se that is at fault, rather the current condition of 'an enduring austerity state' (Jessop, 2016). Jessop views this version of the state as aiming to 'rearticulate relations between (1) the social power of money as capital and of capital as property and (2) the political power of the state' (Jessop, 2016, p. 235). Due to this transformation, less affluent urban citizens are being placed at greater risk of harm than at any time since WWII; their claims for protection and bureaucratic competency from authorities are being unmet.

The right to the city and the values of bureaucracy

This chapter deliberately juxtaposes Lefebvre's humanistic conception of the right to the city with British Standards safety tests, building regulations, ACM cladding panels, polyethylene filler, and Celotex insulation. The point is to expose apparently incompatible differences in language, grammar, and sociological aesthetics; but also to present the necessity, post-Grenfell, of incorporating the right to protection within theoretical and practical articulations of the right to the city. This also points to how there is no need 'to oppose a broadly realist political economy tradition of critical urbanism with a broadly poststructuralist approach to critique' (McFarlane, 2011, p. 204). The radical political potential of an assemblage approach to the right to the city becomes much clearer after Grenfell because the (deadly) consequences of particular interactions between humans and non-human objects are made visible to us. This is also a reminder of the importance of 'recognising the social, cultural, and political power relations embodied relationally in (…) socio-natural or techno-natural imbroglios' (Swyngeodouw, 2006, p. 113). And yet, such a focus on process, materiality and positioning gains more political ballast when it is inserted within avowedly activist discourses such as the right to the city.

For Lefebvre (1996), the right to the city denotes a 'superior right' concerned with inhabiting the city in a profound sense, rather than owning or renting a place to live there or being allowed to work or contribute to decisions made there. It is not a 'right' in a conventional or singular sense. As a Marxist Lefebvre was wary of rights discourse. This is why, as Marcuse argues, Lefebvre is not talking about a legal claim to the city. Rather:

> [It] is multiple rights that are incorporated here: not just one, not just a right to public space, or a right to information and transparency in government, or a right to access to the centre, or a right to this service or that, but the right to a totality, a complexity, in which each of the parts is part of the single whole, to which the right is demanded.
>
> (Marcuse, 2012, p. 35)

The right to the city involves the appropriation of time, space, body, and desire. It prioritises freedom and the ability to enjoy 'individualisation in socialisation' (Lefebvre, 1996, p. 174). It also implies the right to centrality; open access to the centre as a place of sociality, culture, and democracy. The right to the city is at heart a democratic project (see Purcell, 2014), but in Lefebvre's words, democracy is always 'a struggle against the democratic state itself, which tends to (…) become monolithic and to smother the society out of which it develops' (Lefebvre, 2009, p. 61). Appropriation and participation in democratic urban life frees the city from technocratic and capitalist control to become a work of art; an oeuvre that is accomplished under historical conditions. The right to both produce and enjoy the city are – in the spirit of Lefebvre's formulation – integrally linked (Marcuse, 2012, p. 36).

In terms of governance of the city (and of society), Lefebvre favours democratic self-management or *autogestion*. Urban citizenship is understood not as something granted by authority but as a 'dynamic possibility offered to individuals who inscribe themselves into the movement of collectivity, of living together: the city' (Lefebvre, 2009). The roots of *autogestion* are found in the nineteenth century (e.g. Proudhon), but the term gained widespread usage in France during the 1950s following the introduction of the Yugoslav model of workers' self-management, which, in opposition to the centralist Soviet model, was argued to represent an independent route to socialism: a strategy to hasten the 'withering away of the state'. Lefebvre admired the Yugoslav model, viewing it as offering favourable conditions for the development of a less instrumental model of urbanism. As Stanek (2011, p. 234) puts it, Lefebvre was fascinated with self-management as the possibility of the self-production of the individual *within the community but beyond the state*. Self-management, according to Lefebvre, is a struggle against both state and the market, since opposition to one would only give dominance to the other. Moreover, Lefebvre was sceptical regarding the possible 'institutionalisation' of self-management, stressing how it would become an instrument of the state. Lefebvre writes that the '[*autogestion*] cannot escape this brutal obligation: to constitute itself as a power which is not that of the state' (Lefebvre, 2009, p. 147).

Of course, Paris in the late 1960s (when and where Lefebvre wrote *La droit de la ville*) provided a very different context to our own neo-liberal times. As a Marxist (of sorts) in a country dominated at the time by an authoritarian Gaullist state, it was no surprise that Lefebvre urged self-organisation and self-determination among urban communities. He was justifiably wary of any attempt to eradicate dissent, desire, or play from the city. He was worried about the influence of planners, technocrats, and the police. Taking the modern history of Paris into account – the Commune, the massacre of 1961, the 1968 protests, and the *banlieues* uprising of 2005 – Lefebvre is justified in not wanting the state to 'protect' its citizens. Rather, he wanted city dwellers to learn how to do this for themselves. As Purcell puts it

> [What] emerges in Lefebvre's work is a Marxism that rejects the state, that maintains itself as an open and evolving project, and that comes to understand itself as more than anything a democratic project, as a struggle by people to shake off the control of capital and the state in order to manage their affairs for themselves. Thus, his politics can appear at times to be quite a lot closer to anarcho-syndicalism or libertarian socialism than to Marxism.
> (Purcell, 2014, p. 145)

A recent irony is that the rolling back of the state associated with neo-liberal and austerity forms of urbanism (see Brenner & Theodore, 2005; Peck, 2012) encourages, or makes a necessity of democratic self-management. As city dwellers see public services cut and food and energy prices rise, the need to self-manage is become more pressing, resulting in spontaneous, grassroots strategies that attempt to control productive activities. This crisis can be broken down to identify predicaments in relation to energy, housing, work, food, health, education, and travel, and so forth. Innovative self-managed responses to each of these problems are now flourishing in many cities although rarely explicitly linked to the right to the city (see Iveson, 2013; Finn, 2014; Kinder, 2016).

The details leading up to the Grenfell fire, which have been discussed here at length, lend more support to Lefebvre's understanding that the interests of the capitalist state are incompatible with those of the people. But these details and *the connections between these details* (the assemblages of materials, regulations, safety tests, residents, authorities, complaints) are clearly relevant to the right to the city. Inclusion into and oversight over such matters is integral to the battle for the right to the city in our current conjuncture. In fact, Grenfell residents – prior to and after the fire – were *engrossed* in these details. This is because they recognised they had a stake in them; that they were *essential* to their class struggle (and the struggle to stay alive) that is inscribed in urban space (Lefebvre, 1991, p. 55).

Bureaucratic forms are not always purely impositions from above; they can also resonate deeply with cultures, values, and aspirations (Savage, 2005). Paul du Gay (2005) explains how there are now many common-sense assumptions about bureaucracy and personal liberty. The Right (but also sections of the libertarian Left) have come to associate state bureaucracy with 'constraints', viewing bureaucracy as a limit to spontaneous action and transformation. But as du Gay stresses, '[these] constraints are intrinsic to the practice of liberal state administration. They are not by-products that can be removed at will to produce fresher, cleaner, faster, shinier public sector management' (du Gay, 2005, p. 54). He continues,

> [for] instance, values of formal equality of treatment for citizens and due process considerations means that the public administration is constrained in its ability to act 'fast and loose' (…). For the public bureau, the client is everyone and who meets (politically and frequently legally) pre-established criteria for service.
>
> (du Gay, 2005, p. 54)

The Grenfell fire reveals how this 'contract' is now in tatters; how 'due process' has been compromised by decades of neo-liberal reform. The values of bureaucracy must be preserved, du Gay argues, because 'exceedingly tight and "bureaucratic" controls are in place to keep the political system from turning into an instrumentality of private profit for those in its employ' (du Gay, 2005, p. 55). The problem now is that such controls and their enforcement have been weakened, placing the formal equality of treatment for citizens in jeopardy. You may live in the city – you may rent a room, apartment, or house – but still not share the same rights of protection as your wealthier (and whiter) neighbours. As Merrifield puts it, from a Lefebvrian perspective, '[citizen] and city-dweller have been dissociated; what has historically been a core ideal, a core unity, of modern political life has, Lefebvre says, perhaps for the first time, perhaps forever, been wrenched apart, prized open' (Merrifield, 2011, p. 474). The argument presented here is that citizen and city-dweller can only be re-associated by appealing, at least in part, to bureaucratic values. As du Gay concludes, there are, in fact, 'myriad ways in which personal liberty is dependent upon the liberal institutions of state and bureaucracy and not antithetical to them' (du Gay, 2005, p. 55).

From 'community safety' to radical democracy

The problem, post-Grenfell, is how to incorporate the right to protection with the right to the city in a way that engages relevant bureaucratic spheres of the state, but also – in line with the original Lefebvrian ethos – helps to make progress towards more deeply democratic, autonomous forms of urban citizenship. An important concept in critical criminology that is useful in this respect – one which, according to many, was misappropriated by the New Labour government (1997–2010) – is that of 'community safety'. Before this is explored and extended into a discussion of radical democracy, there is a requirement to establish distance between 'safety' and 'security'.

As Neocleous explains, in terms of governmental rhetoric, 'security' now assumes almost a monopolistic character over public discussions of risk and harm. The securitisation agenda, he argues, 'has been coined, shaped and deployed by political, commercial and intellectual forces' (Neocleous, 2008, p. 7). The concern with safety, or more precisely the right to protection stands in opposition to the security-mongering that dominates contemporary politics. In fact, the security agenda – with its support for greater integration of military, police, and cybersecurity apparatuses – falsely conveys the sense that the state is protecting citizens, whereas in many crucial spheres of state activity – such as the management of public housing for example – the right to protection is being eroded with fatal consequences. The illusion created by the discourse of security is powerful. As Vitale argues, the police 'are not here to protect you' (Vitale, 2017, p. 27). The 'security' role they are asked to perform, prevents this. As he explains 'as long as the police are tasked with waging simultaneous wars on drugs, crime, disorder, and terrorism, we will have aggressive and invasive policing that disproportionately criminalises the young, poor, male and non-white' (Vitale, 2017, p. 27). In neo-liberal times, all social problems are turned into police problems. Even the Grenfell Inquiry falls into this category, with its attention focused on *who* or *whom* is guilty and liable for prosecution, rather than addressing the deeper structural issues that allowed the fire to occur and so many people to lose their lives and homes.

Community safety was developed among leftish policy circles during the late 1980s as a critique of the government's prioritising of street crime over and above the other harms that blighted the lives of the urban working-classes. It was presented in the 1991 Morgan Report as an alternative to 'top-down' models of urban policing. The report outlined how the government's concern with 'crime prevention' reinforced the view that the reduction of crime is solely the responsibility of the police. In contrast, community safety was argued to encourage greater participation from all sections of the community including families, victims and 'at risk' groups. Hughes (2007) views community safety as containing the potential for a radical, pan-hazard approach that involved tackling domestic violence, racism, poverty, diet, pollution, mental health, road safety, access to leisure facilities, and so forth on an equal footing to crime. The concern of community safety should be with the harms that disproportionally affect the urban working-class (this is the distinctive hallmark of the left-realist approach to criminology e.g. Lea & Young, 1984). Community safety advocates a horizontal, multi-agency approach to harms that would necessarily be informed by those communities most at risk.

In practice, community safety was sublimated by New Labour into a more conservative and coercive policing concern with 'crime and disorder reduction'. As Gilling (2001) puts it, its understanding of 'unsafety' became restricted to crime-related matters, notably to ameliorating the fear of crime and to punishing anti-social behaviour. It became a normative project, driven by a moral authoritarian brand of communitarianism. Rather than as a radical solution, community safety was utilised as a Third Way policy tool that allowed neo-liberal reforms to be accompanied by warm rhetorical tones about the importance and value of community. Gilling argues that what community safety addressed in practice was

> the insecurity not of the victims of neo-liberalism (the socially dislocated underclass), but of the 'respectable classes', whose enhanced freedom and affluence has come at the expense of growing insecurity about the dangerous other.
>
> (Gilling, 2001, p. 397)

Radical potential in a 'pan-hazard' approach to community safety remains. As Little (2010) argues, communities are domains where neo-liberal economic rationality need not prevail. Little aligns with Mouffe (1992) and backs a position of radical democracy. The problem with Third Way mobilisations of community, he argues, is that they deny complexity and de-politicise differences and conflicts (Little, 2010, p. 374). Its advocacy of community is an attempt to create 'a politics without adversary' and to 'ignore social divisions and inequalities of power' (Little, 2010, p. 374). Radical democracy, in contrast, insists on the assumption of value pluralism; that there are a range of communities in any given society. Such communities will inevitably express a plurality of non-fixed viewpoints. Sometimes there will be irreconcilable differences between viewpoints this disagreement can be used to deepen engagement (Little, 2010, p. 376). Little also argues that 'the recognition of different communities needs to be coupled with a strategy for redistribution that can help to empower those communities that have been socially excluded or marginalised from power and decision-making' (Little, 2010, p. 376). This involves the establishment or radical transformation of political institutions so that 'a greater range of voices can be heard in a wider range of locations than is the case within the narrow parameters of contemporary liberal democratic politics' (Little, 2010, p. 377). This goes beyond the state simply 'listening'; instead it involves the opening-up of political spaces and bureaucracies to the degree where say, in the wake of Grenfell, social housing residents choose refurbishment plans and contractors and even be trained to take part in the work, are invited as lay witnesses to fire tests of building materials, and active participants in agreeing local authority budgets for essential services such as firefighting. As Mouffe puts it, 'the real task (...) is to foster allegiance to our democratic institutions (...) by creating strong forms of identification with them' (Mouffe, cited in Little, 2010, p. 278). Yet, radical democracy only occurs through the state empowering communities via economic redistribution 'whilst [also] recognising that *some elements of democratic ethos can extend beyond the walls of the state*' (Little, 2010, p. 380). In such a scenario, the right to the city is supported by the state but is claimed (and lived) independently from the state. The crucial aspect, which chimes with the openness of Lefebvre's notion of the right to the city, is 'the impossibility [and undesirability] of (...) a perfectionist scenario emerging' (Little, 2010, p. 380).

A 'perfectionist scenario' in relation to community safety would be total safety for all urban residents which, of course, is an absurd proposition. 'Safety' is an absolute and therefore less compatible with the notion of the right to the city as is the right to *protection*. Moreover, total safety is not desirable in cities that thrive on risk and encounter; indeed, safety may be read as antithetical to freedom. As Sennett points out in *The Uses of Disorder*, cities are a way of teaching people to live with openness, freedom, and conflict; it is impossible to make such a milieu 'safe' and nor should we want to. 'Making peace', he argues, is a responsibility for residents themselves, not the police or the state (Sennett, 1970, p. 164). An emphasis on protection also avoids paternalistic interpretations of 'safety', which imagines urban dwellers – especially women – as vulnerable, passive and in need of rescuing (see Wilson, 1992). The upshot is that women (and other 'vulnerable' groups) become excluded from full participation in urban life (the very opposite of the right to the city).

A transformation of community safety into the right to protection – under the rubric of radical democracy – has the potential to identify, act upon, and neutralise the hazards of urban living that have been created and exacerbated by decades of rolling back the remit and influence of the state. Its pluralist force can also be used to foreground issues of misogyny, abuse, hate crimes, pollution, poor health, and animal cruelty; again, issues upon which a Lefebvrian conception of the right to the city remains silent. On one hand, it will engage or revive bureaucratic activities of the state with which residents have a stake, such as building control; but on the other, it can deepen the radical conception of the self-determined urban citizen that Lefebvre believes the right to the city rests upon. Yet, deep democratisation must tread carefully. Lefebvre (2009) is right to be wary of how 'organic' self-management strategies can easily be co-opted by the state to further different ends, a growing problem under conditions of neoliberal urbanism. And, as Mark Purcell puts it in *The Down-Deep Delight of Democracy*, 'democratic connections must help create some sort of solidarity, commonality, and consistency among people. But they must also be such that they ward off the formation of oligarchy and heteronomy' (Purcell, 2013, p. 124). He argues the task is not to *make* democracy happen, but to *search for it*, and to *recognise* the participation and eagerness-to-participate that already exists. It is within this locus of 'the struggle to become democratic' (Purcell, 2013, p. 157) where we find the Grenfell Action Group; a community of engaged citizens claiming their right to protection and their right to the city, but prior to Grenfell going *unrecognised*, being *ignored*, and treated with the *contempt* reserved for conspiracy-theorists (see O'Hagan, 2018). Such groups of citizens of course exist *everywhere* (there are now many more in existence who are connected to Grenfell Tower); their actions and transformative demands can be found in every housing estate, ghetto or barrio across the globe.

What is envisaged here, the result of a sometimes awkward trajectory through community safety and radical democracy is the right to protection, a classic Lefebvrian 'virtual object (...); that is, a possible object, whose growth and development can be analyzed in relation to a process and a praxis (practical activity)' (Lefebvre, 2003, p. 3). For Lefebvre, a virtual object is not a utopia, an ideal that could never exist. Rather it is an extrapolation or amplification in thought of practices and ideas that are *already* taking place in the city; practices and ideas that have not yet come to full maturity. It involves recognition for the communities who are already immersed with the facticity of materials, tests, bureaucracies and so forth but also springs from a humanistic source, from 'the politics of sympathy discernible in the shadows of disaster' (Gilroy, 2019).

Aesthetics

According to the cultural theorist Virilio (2007), the 'accident' reveals much that had previously been hidden. The accident, he argues, contradicts its own apparent uniqueness or exceptionality to disclose a submerged world. The submerged world revealed by Grenfell are all the pre-Grenfell fires that can be attributed to combustible plastic cladding such as Garnock Court, Harrow Court, Lakanal House, Shirley Towers, and Shepherd's Court (all in the UK since 1999), as well as 'the other Grenfells in waiting', the many other residential tower blocks decked out in highly flammable cladding (Hodkinson, 2019, p. xi). As highlighted in the preceding sections, materials such as cladding and fillers, laboratory tests, regulations, fire brigade budgets are integral to the right to protection; they have been made visible by the Grenfell fire. Virilio (2007) intimates that events influence aesthetics in that they present something (or some things) to 'represent'. If we decide to persevere with the notion of the right to the city its success will, in part, be due to whether or not a new aesthetics of the right to the city can be forged in the context of Grenfell. Rancière (2004) also makes clear the relation between aesthetics and politics, pointing to how what can be seen and what can be counted denotes the domain of the 'sensible'. Alone, aesthetics do not constitute transformative politics (Rancière, 2009), but radicalism only becomes an option with the ability to *see* 'the submerged' and/or 'the possible'; or is realised via an aesthetic regime that while not wholly 'representative' (of an accident that is impossible to represent through language) still offers a judgement on appropriate and inappropriate means of representation (Rancière, 2007, p. 126). The silent marches that take place on the evening of the 14th of each month, which wind through residential streets close to the tower and which act both as memorial and protest can be read as an at attempt at a new aesthetics of the right to the city.

The aesthetics of the right to the city has so far been dominated by the image of dense, noisy, carnivalesque occupations of public space by protestors. It is a centripetal vision, a drawing of people towards the appropriation of an urban centre. This is largely derived from Lefebvre's (1996) arguments that the right to the city lies in the restitution of use value from the clutches of exchange value and the growing predominance of capitalist or 'abstract space'. Take this description, a 'caricature' of Lefebvre's ideal from David Harvey's *Rebel Cities*:

> [The] spontaneous coming together in a moment of 'irruption', when disparate heterotopic groups suddenly see, if only for a fleeting moment, the possibilities of collective action to create something radically different.
> (Harvey, 2012, p. xvii)

This draws upon the irreverent spirit of situationism and the ethos of Paris May 1968; it even contains an impulse for the resurrection of such times, spaces, and practices. Another aesthetic dimension of the right to the city to gain traction after the Arab Spring and the anti-capitalist occupations of 2011 is that of 'horizontalidad', an attempt to construct non-hierarchical forms of occupation, communication, and habitat (see Burgum, 2018).

The Grenfell fire necessitates a shift from this distribution of the sensible, an awakening of the 'optical unconscious' (Benjamin, 1999) to include building materials, legal documents, laboratories, etc. in addition to public squares, demonstrations, and occupations. In much the same way, Larkin (2013) argues that it is not only use value that matters when it comes to urban infrastructures (and materials) but also *what they signify*, what they *stand for* in terms of the relation of trust between citizens and the state (see also Garbin & Millington, 2018). As such, it is necessary to recognise bureaucracies, aluminium composite materials, and sprinkler systems as aspects of – or *stakes in* – an urban world that can, despite the esoteric nature of these entities, be politicised *from below;* mobilised as symbolic resources in struggles for the right to the city. Such an aesthetic is more attuned to our times: less humanistic (without abandoning humanism entirely) but more inclusive, sober and directed more towards details, process and the *longue durée* rather than *derive* and the 'moment'.

Conclusion

Henri Lefebvre's notion of the right to the city has not been as helpful as one might expect following the catastrophic Grenfell Tower fire in London, in June 2017. Despite enjoying a renaissance in twenty-first-century scholarship and activism there is a real prospect that the right to the city has lost relevance in post-crash, austerity-era cities where out of necessity everyday working-class struggles prioritise obtaining safe housing, finding secure employment, opposing police racism, and avoiding hunger. One problem is Lefebvre's refusal to grant any role in the right to the city to bureaucracy and technocracy. It is suggested that in neo-liberal times such as ours, urban residents – especially those from poorer communities – appeal to the *values* of bureaucracy (rather than actually existing bureaucracies) in order to ensure their homes are safe. Even if this is somewhat at odds with Lefebvre's formulation, the right to protection has become an integral aspect of the right to the city in neo-liberal cities, threatening even to override concerns with freedom, playfulness, and creativity. The chapter began by analysing various technically-focused reports in relation to the Grenfell fire. These were mainly concerned with the flammability of the building materials used in the refurbishment of Grenfell Tower in the years immediately preceding the fire. These reports examined regulations, materials, testing procedures, and enforcement, pointing to many failures caused by attempts to remove or 'streamline' regulation and local authority austerity measures that led to cost-cutting. The chapter then returned to the right to the city, focusing on Lefebvre's preference for self-determination over the authority of the state. The point here was to place Lefebvre's writing in context and to show how this context has changed drastically in a half-century of political and economic transformation. Lefebvre's hostility towards the state – which is not misplaced – is juxtaposed with Paul du Gay's assertions that (1) bureaucracy has been compromised by political attacks on 'red tape' leading to reductions in safety and fairness; and (2) state bureaucracy is a condition of freedom. Finally, the chapter considers whether the left-realist inspired policy notion of community safety can be aligned with more radical forms of democracy to reanimate political spaces and bureaucracies in the struggle for the right to protection. It is argued that such an alignment helpfully finds some affinity with Lefebvre's right to the city, especially with regard to its stresses on deepening democracy and promoting self-management. This argument – in working towards incorporating protection into the rubric of the right to the city – holds left-realism and left-idealism in productive tension. It calls for a deliberative strategy of moving between these two poles in the context of mobilising pluralistic urban communities post-Grenfell in addition to an aesthetic of the right to the city that is extended beyond the tropes of festival and occupation. In terms of critical urban studies and citizenship studies, future work that is grounded in Marxist political economy but which also recognises the assemblages of humans and non-humans (e.g. documents, materials, gases) that comprise the domain of how the social (and urban) is enacted (Law & Urry, 2004) is encouraged; more specifically, studies that enable an understanding of how the right to protection is denied and/or the ontological and aesthetic grounds upon which it might be successfully claimed in the future.

References

Apps, P., Barnes S., & Barratt, L. (2017). *The paper trail: The failure of building regulations.* Retrieved from https://social.shorthand.com/insidehousing/3CWytp9tQj/the-failure-of-building-regulations-thepaper-trail.

Atkinson, R., & Millington, G. (2019). *Urban criminology.* London: Routledge.

Benjamin, W. (1999). *The Arcades Project.* Cambridge, MA: Harvard University Press.

Bisby, L. (2018). *Grenfell tower inquiry phase 1 – final expert report.* Edinburgh: University of Edinburgh School of Engineering. Retrieved from www.grenfelltowerinquiry.org.uk/evidence/exhibits-professor-luke-bisbys-expert-report.

Bodnar, J. (2013). What's left of the right to the city? In D. J. Sherman, R. van Dijk, J. Alinder, & A. Aneesh (Eds.), *The long 1968* (pp. 73–90). Bloomington, IN: Indiana University Press.

Booth, R., Gentleman, A., & Khilali, M. (2017). Grenfell Tower gas pipes left exposed, despite fire safety expert's orders. *Guardian*, 27 June. Retrieved from www.theguardian.com/uk-news/2017/jun/27/grenfell-tower-gas-pipes-left-exposed-despite-fire-safety-experts-orders.

Boughton, J. (2018). *Municipal dreams: The rise and fall of council housing.* London: Verso.

Brenner, N., & Theodore, N. (2005). Neoliberalism and the urban condition. *City, 9*(1), 101–107.

Bulley, D., Edkins, J., & El-Enany, N. (Eds.) (2019). *After Grenfell: Violence, resistance and response.* London: Pluto.

Burgum, S. (2018). *Occupying London: Post-crash resistance and the limits of possibility.* London: Routledge.

Chakraborrty, A. (2017). Over 170 years after Engels, Britain is still a country that murders its poor. *Guardian*, 20 June. Retrieved from www.theguardian.com/commentisfree/2017/jun/20/engels-britain-murders-poor-grenfell-tower.

Cooper, V., & Whyte, D. (Eds.) (2017). *The violence of austerity.* London: Pluto Press.

du Gay, Paul (Ed.) (1996). *The values of bureaucracy.* Oxford: Oxford University Press.

Finn, D. (2014). DIY urbanism: implications for cities. *Journal of Urbanism: International Research on Placemaking and Urban Sustainability, 7*(4), 381–398.

Garbin, D., and Millington, G. (2018) 'Central London under siege': Diaspora, 'race' and the right to the (global) city, *The Sociological Review, 66*(1), 138–154.

Gilling, D. (2001). Community safety and social policy. *European Journal on Criminal Policy and Research, 9*(4), 381–400.

Gilroy, P. (2019). Never again: Refusing race and salvaging the human. *New Frame*, 20 June 2019. Retrieved from www.newframe.com/long-read-refusing-race-and-salvaging-the-human/.

GAG – Grenfell Action Group (2016). *KCTMO – Playing with fire!* Retrieved from https://grenfellactiongroup.wordpress.com/2016/11/20/kctmo-playing-with-fire/.

Hackitt, J. (2018). *Building a safer future: Independent review of building regulations and fire safety final report.* London: Secretary of State for Housing, Communities and Local Government.

Harvey, D. (2012). *Rebel cities.* New York, NY: Verso.

Hodkinson, S. (2019). *Safe as houses: Private greed, political negligence and housing policy after Grenfell.* Manchester: Manchester University Press.

Hughes, G. (2007). *The politics of crime and community*. Basingstoke: Palgrave Macmillan.
Iveson, K. (2013). Cities within the city: Do-it-yourself urbanism and the right to the city. *International Journal of Urban and Regional Research, 37*(3), 941–956.
Jessop, B. (2016). *The state: Past, present, future*. Cambridge: Polity.
Kentish, B. (2017). Grenfell Tower fire: Theresa May apologises for 'failure of the State' after widely criticised response, *Independent*. Retrieved from www.independent.co.uk/news/uk/home-news/grenfell-tower-fire-theresa-may-apology-response-government-failure-state-latest-news-a7801246.html.
Kinder, K. (2016). *DIY Detroit: Making do in a city without services*. Minneapolis, MN: University of Minnesota Press.
Kofman, E., & Lebas, E. (1996). Lost in transposition – time, space and the city. In H. Lefevre, *Writings on cities* (pp. 3–63). Oxford: Blackwell.
Larkin, B. (2013). The politics and poetics of infrastructure. *Annual Review of Anthropology, 42*, 327–343.
Law, J., & Urry, J. (2004). Enacting the social. *Economy and Society, 33*(3), 390–410.
Lea, J., & Young, J. (1984). *What is to be done about law and order?* London: Pluto Press.
Lefebvre, H. (1996). *Writings on cities*. Translated and edited by Eleonore Kofman and Elizabeth Lebas. Cambridge, MA: Blackwell.
Lefebvre, H. (2000). *Everyday life in the modern world*. London: Continuum.
Lefebvre, H. (2003). *Key writings*. London: Bloomsbury.
Lefebvre, H. (2009). *State, space, world: Selected essays*. Minneapolis, MN: University of Minnesota Press.
Little, A. (2010). Community and radical democracy. *Journal of Political Ideologies, 7*(3), 369–382.
MacLeod, G. (2018). The Grenfell Tower atrocity: Exposing urban worlds of inequality, injustice, and an impaired democracy. *City, 22*(4), 460–489.
Madden, D. (2017). Deadly cityscapes of inequality. Retrieved from www.thesociologicalreview.com/blog/deadly-cityscapes-of-inequality.html.
Madden, D., Marcuse, P. (2016). *In defense of housing: The politics of crisis*. London: Verso.
Marcuse, P. (2012). Whose right(s) to what city? In N. Brenner, P. Marcuse & M. Mayer (Eds.), *Cities for people, not for profit* (pp. 24–41). New York, NY: Routledge.
McFarlane, C. (2011). Assemblage and critical urbanism. *City, 15*(2), 204–224.
Merrifield, A. (2002). *MetroMarxism*. New York, NY: Routledge.
Merrifield, A. (2011). The right to the city and beyond: Notes on a Lefebvrian re-conceptualization. *City, 15*(3–4), 473–481.
Mouffe, C. (1992). *Dimensions of radical democracy: Pluralism, citizenship, community*. London: Verso.
Neocleous, M. (2008). *Critique of security*. Edinburgh: Edinburgh University Press.
O'Hagan, A. (2018). The tower. *London Review of Books*. www.lrb.co.uk/v40/n11/andrew-ohagan/the-tower.
O'Hara, M. (2015). *Austerity bites: A journey to the sharp end of the cuts in the UK*. Bristol: Policy Press.
Peck, J. (2012). Austerity urbanism: American cities under extreme economy. *City, 16*(6), 626–655.
Platt, E. (2017). Grenfell Tower: Chronicle of a tragedy foretold. *New Statesman*, 9 October. Retrieved from www.newstatesman.com/politics/uk/2017/10/grenfell-tower-chronicle-tragedy-foretold.
Purcell, M. (2013). *The down-deep delight of democracy*. New York, NY: Wiley.

Purcell, M. (2014). Possible worlds: Henri Lefebvre and the right to the city. *Journal of Urban Affairs*, *36*(1), 141–154.
Rancière, J. (2004). *The politics of aesthetics*. London: Bloomsbury.
Rancière, J. (2007). *The future of the image*. London: Verso.
Rancière, J. (2009). *The emancipated spectator*. London: Verso.
Renwick, D. (2019). Organising on mute. In D. Bulley, J. Edkins & N. El-Enany (Eds.), *After Grenfell: Violence, Resistance and Response* (pp. 19–46). London: Pluto.
Savage, M. (2005). The popularity of bureaucracy: Involvement in voluntary associations. In P. du Gay (Ed.) *The values of bureaucracy* (pp. 309–335). Oxford: Oxford University Press.
Sennett, R. (1970). *The uses of disorder: Personal identity and city life*. Harmondsworth: Penguin.
Shilliam, R. (2019). The fire and the academy. In D. Bulley, J. Edkins & N. El-Enany (Eds.), *After Grenfell: Violence, resistance and response* (pp. 195–197). London: Pluto.
Stanek, L. (2011). *Henri Lefebvre on space*. Minneapolis, MN: University of Minnesota Press.
Swyngedouw, E. (2006). Circulations and metabolisms: (Hybrid) natures and (cyborg) cities, *Science as Culture*, *15*(2), 105–121.
Tombs, S. (2017). Undoing social protection. In V. Cooper, & D. Whyte, (Eds.), *The violence of austerity*. (pp. 133–140). London: Pluto Press.
Virilio, P. (2007). *The original accident*. Cambridge: Polity.
Vitale, A. (2017). *The end of policing*. New York, NY: Verso.
Wilson, E. (1992). *The sphinx in the city: Urban life, the control of disorder, and women*. Berkeley, CA: University of California Press.

Index

activating society, concept of 88
advanced marginality 21–2, 24; schema of 25; sociology of 17
African Americans 21; urbanites 19
anti-ghettos 21, 24; of Western Europe 25
Arab Spring (2011) 147
assured shorthold tenancies (AST) 105–6
austerity politics: conditions of 7; consequences of 5, 10, 120; and crisis in place 120–1; financial assistance programmes 120; in Greece 119; precarious lives in persisting austerity 125–6; result of 122

banks 2, 4, 9, 39, 41–2, 119, 120, 127
banlieues uprising of 2005 141
Berlin 10, 78–94, 100–1, 107–8, 110, 113
black marginality, political roots of 18–19
Black Metropolis 19
black middle-class districts (BMCD) 19
bureaucracy, values of 148
bureaucratic society, of controlled consumption 132

capitalism 3, 22, 45, 58, 83, 88, 132
capitalist: capital accumulation 32, 34; capital switching 108; capital investment 75; crises theorisations 44; flexible capitalism 88; political economy 100; valorisation, dynamics of 91
Charter Cities 6, 50–3
cities-as-businesses 6
cities, successful 4–5, 7, 76, 79
citizenship: citizen-as-consumer 3; contract 54; entitlements of 2; experiences of neo-liberal attacks 10; political concept of 50; rights of 2, 11, 102; second-class citizen 113; social citizenship 120; urban 3, 139, 141, 143, 145

civil society 8, 78–9, 81–2, 84–5, 92–3
class: creative class, idea of 42; dangerous 24, 39, 43; dissolution 23; struggle 3, 142, 148
colonial 35, 65
commodification of city 3–6
community safety, notion of 133, 143–6, 148
competition: international 63, 69; power of 53; principle of 3–6, 8, 50
construction industry 69, 136, 139
Corporate Social Responsibility 78
crisis: banking 41; financial 2, 83, 135; immigration 65

deindustrialisation: effects of 64; issue of 65
democracy 2–3; liberal 8–9, 145; participation 3–4, 6, 8, 52, 54, 56; radical 143–6; rise and development of 5; western 51
democratic self-management 141
de-municipalisation, process of 134
deregulation, doctrine of 12, 26, 52, 100, 103–8, 114
development strategies, outsourcing of 85, 93
displacement anxiety, emotional aspects of 110–11, 114
displacement, risk of 2, 91, 106–7, 109, 111, 114

economic restructuring, effects of 63–4, 68, 73
education 2, 19, 31, 44, 69–70, 72, 81, 87, 120, 125, 134, 142
elites: financial 9; global 2, 5; global financial 2, 9; neo-liberal global 2, 69, 78
empire 34–5, 43

era, post-industrial 66, 74–6
ethnic identity, role of 21
ethnicity, definition of 21
European Union (EU) 85, 119; attitudes on refugees 125
European working-class periphery 17, 20, 23
exclusion 9, 21, 78, 89–90, 93, 102, 121; social 9, 80, 91, 113
experience, urban 8–11

financial crisis of 2007–2008 2, 83, 135
financialisation 44, 103, 108; of housing 114; meaning of 104
financial securitisation 106
free-market economy 50, 52, 57
Free Private Cities 6, 53–5, 56, 58

gecekondus 37, 45
gentrification 68, 83, 91, 108
geography, urban 8
ghetto 18–19; black American ghetto 23; *cité-ghetto* 23; *ghetto-område* 23; hyperghetto (HyGh) 18–19; sink estate 23
ghettoisation: discourse of 19; hyper-ghettoisation 43; paradoxical profits of 24
Global Cities 4, 39, 70
globalisation 31, 63, 66–7, 78
governing the city: citizens' participation in 8; principle of competition 6; uneven developments and institutional logics 6–8
governmentality, concept of 8, 83, 86, 89, 92
grassroots movements 39, 83
Greece 10, 119–20, 127
Grenfell Action Group (GAG) 134, 146

health care and social assistance 63, 73
health system 52
hedge-funds 2, 4, 9, 12
home: concept of 102; under conditions of neo-liberalised housing 108–13; multifacetedness and multiscalarity of 102
homelessness 99, 101, 106, 113, 120
homeless people 99, 121
homeownership: in Berlin 107; desire for 104; in Greece 123; promotion through subsidies 108; rates of 104; and rise in real estate prices 105
housing: access and entitlement to 102; as commodity 102; crisis 2, 10, 99, 100; in Greece 123; investments 37; market 37–8, 40, 44, 91, 101, 107;

neo-liberalisation of *see* neo-liberalised housing; privatisation 104; right to 102; Right to Let 105; social 11, 41–2, 99, 104–5, 107, 110, 135, 145; 'Southern' model of 123; theory of 101–3; urban neo-liberalism 103; and urban planning 40; use of 102
housing crisis: in Berlin 107–8; *Housing Crisis, The* 99–100; in London 104–6; no fault eviction 105; occurrence of 100; renting insecurity 103–8; revenge evictions 105
housing emergencies, quality of 101
housing industry, financing of 40
housing insecurity: owner-occupation and renting 101; planning and building guidelines 100; repossessions of rented homes by landlords 101; social problems and 101; and sub-standard living 99
housing subsidies, reduction of 38, 103

ideology 1, 4, 10, 33, 53–4, 103
immigrants 64–5, 67, 72
immigration 18, 64–5
income inequalities 121, 123
industrialisation 3, 37
inequality 18, 21, 24–6, 101, 112, 135; income 121–3; social 3, 10–11, 78, 90, 93, 104
inhabitance, notion of 3
inner city 18, 38–9, 42–3, 63, 65, 68–73, 93, 107
inner-city living, lifestyle opportunities of 66
inner-city neighbourhoods 39, 43, 72
insecurity 10, 128, 144; of home 113; job 126; renting 101, 103–8, 114; social insecurity 22
investment 2–4, 7, 9–12, 37, 66, 71, 75, 79, 81, 88, 105–8, 114, 120, 128
Istanbul 30–45

job: insecurity 126; losses 73; opportunities 124
joint ventures 69

landlords: generation landlord 105; power imbalances with renters 112; private 105; repossessions of rented homes by 101; small-time 105
liberal democracy 8–9
life-chances, accumulation of 8
lifestyle 43–4, 66, 89

liminality, idea of 10, 23, 110, 114
London, housing crisis in 104–6; assured shorthold tenancies (AST) 105–6; *Buy to Let* mortgages (BTL) 106; neo-liberalisation of housing 106; private rental sector (PRS) 100, 106; Right to Buy scheme (RTB) 104–5; Right to Let 105

mahalle 39–40; historical hagiography of 43; physical disappearance of 44; slum *mahalles* 44
managers 12, 64
marginality 17–18, 21–6
market freedom, idea of 6
marketisation 2, 12
mass immigration 65
mass unemployment 69
material deprivation, index of 18, 121, 129n2
Melbourne 63–76
metropolis: American 18–19; Black 19
metropolitan 25, 63–73, 120–1, 125
modernisation 30, 35–6, 107–8
mortgages 37, 123; finance 103; market 106
municipal administration, responsibilities of 88
municipal companies 81
municipal housing associations 81, 88, 90, 92–3
municipal-sponsored initiative 74

neighbourhood development 92; cooperation with municipal housing association 87–8, 90; relationship with gentrification 91
neighbourhood effects 22
neighbourhood management 80, 81, 84–7, 89; conditions for participation and activism 89–92; guidelines in 90; *see also* Berlin's neighbourhoods
neo-liberal cities 1, 78–9, 90, 148
neo-liberalised housing: emotional aspects of displacement anxiety 111; existential insecurity 110; home under conditions of 108–13; injustice and guilt 113; moral 113; power and agency 112; violation of privacy 111–12
neo-liberalism: definition of 1; soft neo-liberalism 87; urban 1–2, 9, 11, 103; war against democracy and citizens 11
neo-liberal politics 39; of asymmetrically distributed life-chances 8–11
neo-liberal restructuration projects 1, 78

neo-liberal urban development processes 78, 82
networks 6, 17, 30, 33–4, 50, 55, 57, 71, 84, 126
non-profit organisations 84, 92

Ottoman Empire: central bureaucracy 35; *mahalle* 35–6; *millet* system 35; territorial breaking up of 35

pensions 7, 52, 106, 124; cuts 119, 122–3; disability 73; loss of 5; national 124; private 108; reduction of 120
planning and building guidelines, deregulation of 100
political-economic transformations 100
political reality, problems of 56
populism 31, 45
post-Fordist restructuring processes 72
post-industrial Europe 19
poverty: anti-poverty policies 86; child poverty 85
precariat 18, 22–5, 109
privacy, violation of 111–12
private rental sector (PRS) 100–1, 106
private renters, living in poverty 101, 105, 108–9, 114
private renting, tenure of 100–1, 109–11, 113
privatisation: asset stripping and 120; free-market policies of 135; issue of 103
profit-maximising, strategies of 2
property: buying and re-selling of 107; taxation on 124
property-owning democracy 104, 114
public housing 21, 23, 42, 71, 104, 134; construction and maintenance of 107; management of 143; and rent control 103; residential buildings 134
public ownership of land 39
public-private partnerships 78, 83

quality of life 78–9, 81, 85, 88–9, 93, 134

radical democracy 143–6
real estate: funds 108; markets 4; prices 44, 105
refugees: and asylum seekers 74; EU attitudes on 125
regulation 10–12, 26, 52–5, 78–9, 92–3, 136, 139, 148
relegation, urban 5, 17, 25
religion 11–12, 54, 58, 65
religious identities 45, 65

rental contracts, termination of 101, 109
rent control 82, 103; abolishment of 105; and 'regulated tenancies' 105
renting: and assured shorthold tenancies (AST) 105; generation rent 105; insecurity through urban neo-liberalism 103–8; owner-occupation and 101; repossessions of rented homes by landlords 101; Right to Let 105
residential segregation 38, 44
'responsibilisation' of citizens 8, 10, 79, 88–9, 92–3, 114
right to the city: aesthetics of 146–7; notion of 3, 11, 131–3, 148; and values of bureaucracy 140–3
'roll back' movement 104

security 9, 11–12, 51, 54, 64, 102–3, 105, 107–9, 125, 127, 143
self-determination, sense of 54, 133, 148
self-employment 120
self-responsibility 9–10, 86
Social City 11–13, 80
social cleansing 9, 135
social exclusion, boundary of 9, 80, 91, 113
social housing 41, 104; protection for residents of 135; scarcity of 105
social inclusion 78, 89, 93
social inequality 3, 11, 93, 104, 112
social insecurity 25; spread and normalisation of 22
social insurance 106
Socially Integrative City programme 79, 81, 85, 88, 91
social market economy 12
social murder, notion of 135–6
social relations 3, 11–12, 21, 25, 32, 45, 102
social responsibility 78, 81
social rights, erosion of 44, 104
social security 111, 122; destruction of 11; in Greece 123; payment of 123; reduction of 108
social space: in slum *mahalles* 44; in TPSN framework 32–40
social violence 109, 121, 139
social welfare state 80, 90
society: industrial 4, 42, 65; liberal-democratic 9; post-industrial 65
socio-spatial inequalities 107, 119
solidarity, obligation of 90
sovereignty 5, 51
space 3; production of 30; urban space *see* urban space

spatial segregation 8; ethnic and faith-based 36
spatialities 34, 128
speculation 2–5, 9, 41–2, 44, 108, 113
state bureaucracy, power of 38, 132, 142, 148
state-free territory 51
statehood, idea of 32
state, local 87, 89, 92, 135
state power, geo-historical periodisation of 34
state-run abilities, decline of 79, 93
state-society relations 87, 89, 92
state-spaces: and emergence of unhindered boosterism 40–4; modalities of 36, 39; territorial expansion of 37
state spatiality, idea of 6, 32
successful cities 4, 79; global elite's strategy of 5; idea of 7; as strategy to de-politicise cities 5; vision of 5
symbolic power 17, 25, 30, 39, 42; Bourdieu's theory of 23

tenants, stigmatisation of 104
tenure, insecurity of 109
territorial stigmatisation 23–5
territories: idea of 33; of poverty 17
territory, place, scale, and network (TPSN) framework: etatist/national developmentalist state-spaces 36–9; formulating the 'social space' in 32–40; mode I of 34–6; mode II of 36–9; networks 34; place 33–4; scale 34; in sociological study of urban matters 33; strategic-relational approach 33; territory 33; world-imperial-quasi-colonial state-spaces (1839–1920s) 34–6
Thatcherism 133
trade, retail 63, 73
troika 119
Turkish identity 44
Turkish state: central planning authorities of 37; characteristics of 36; financial and cultural prowess 42; Housing and Urban Development Act (1984) 40; Housing and Urban Development Fund (TOKI) 40–2; Islamic identity 42; Kemalist national developmentalism 42; migrant *mahalles* 39, 42; military interventions of 1980 and 1997 39; movie industry 38; non-Turkified minorities 36; popular music 38; state and society relations in 36; white Turks 44; *see also* Istanbul (Turkey)

unemployment rate 22, 67, 73, 119
Universal Declaration of Human Rights 102
urban citizenship 141; forms of 143
urban governance 84, 85; rationality of 86–7; through community 86–7
urban grassroots movements 83
urban housing: development 42; precarity 109
urban inequality, sociology of 24–6
urban life, social and cultural benefits of 132
urban marginality: regime of 21, 26; and state-crafting 26
urban neo-liberalism: concept of 1; ideology of 103; processes of 1; renting insecurity through 103–8
urban outcasts 18; convergence thesis of 19–21; emergence thesis of 21–4; of foreign origins 21

urban sociology 5, 25, 31, 80; schools of 31
urban space 128, 142; concept of 31; in Istanbul 30–2; living in and creating 3; transformation of 1; use of 3

wage labour, fragmentation of 22
welfare state 8, 11, 79–80; asset-based welfare provision 108; dismantling of 108, 114; financial instruments 106; Fordist 82; framework for shifting 88; as 'neo-social' arrangement 89; responsibility of 78; social 80, 90; transformation of 88
welfare system, asset- or property-based 106
World Economic Forum 3, 4